'This is a soaring, joyful book, filled with the wit and wonder of aerial gymnastics, deep time, evolution and biology. It might just be the nearest thing to flight in a literary form.'

Patrick Barkham, author of *Wild Green Wonders*

'Lev Parikian illuminates one of the most magical mysteries of the natural world: how birds, bats, and insects break the bounds of Earth and fly through the heavens. His prose is every bit as buoyant as his airborne subjects. With the conversational chops of a storyteller and the authoritative expertise of a scientist, Parikian is a nature writer at the top of his game.'

Steve Brusatte, author of *The Rise and Fall of the Dinosaurs*

'Witty and enlightening. This book may not give you wings, but it will give you a deep appreciation for all those animals that glide, soar, hover and flutter . . . and penguins.'

Helen Pilcher, author of *Life Changing*

'A beautiful concept, flawlessly executed, *Taking Flight* is among the most charming popular science books I've read in years. Parikian is fast becoming one of the finest science writers out there.'

Jules Howard, author of *Wonderdog*

'Lev Parikian explores one of nature's most astounding evolutionary conjuring tricks . . . A work of clarity, levity and joy.'

Caspar Henderson, author of *A New Map of Wonders*

'Whether you're an engineer, a linguist, a historian or just curious, this book has something for you. It explodes with fascinating titbits, all delivered with a gentle touch of humour. Get ready to be swept away by the expertly crafted harmony of *Taking Flight*.'

Professor Lucy Rogers, author of *It's ONLY Rocket Science*

'Lev Parikian has produced a clear, crisp and entertaining account of the history of animal flight. A delightful and insightful read.'

Dominic Couzens, author of *A Bird a Day*

'Lev Parikian's writing – about the extraordinary wonders of flight – is as magical and uplifting as the aerial dynamics of our tiniest insects and birds.'

Ann Pettifor, author of *The Case for the New Green Deal*

'Had P. G. Wodehouse ghost-written Attenborough's *Life on Earth*, we might have had *Taking Flight* forty years ago. This is a charming book, which – like its author – fizzes with erudition, wordplay and humour.'

Nick Acheson, author of *The Meaning of Geese*

'*Taking Flight* is full of wonders, large and small, and Lev's own sense of the astounding fact of flight will make you look at the world differently. Those of us who can't distinguish a pigeon from a pterosaur will learn lots, but it's also bang up to date and informed by research. Brilliant stuff – I'm already planning to press copies on friends.'

Chris Lintott, University of Oxford

'This book is fascinating – packed with "well, I never!" and "who'd have thought?" lines which you feel compelled to share. Beautifully written with characteristic warmth and humour, Lev somehow manages to explain the mechanics of flight in glorious detail without dispelling the miracle and magic of such a feat.'

Brigit Strawbridge, author of *Dancing with Bees*

TAKING FLIGHT

TAKING FLIGHT

The evolutionary story of life on the wing

LEV PARIKIAN

Elliott&Thompson

First published 2023 by
Elliott and Thompson Limited
2 John Street
London WC1N 2ES
www.eandtbooks.com

ISBN: 978-1-78396-703-2

9 8 7 6 5 4 3 2 1

A catalogue record for this book is available from the British Library.

Typesetting by Marie Doherty
Printed by CPI Group (UK) Ltd, Croydon, CR0 4YY

To Winter.

Cat of cats, occasional keyboard sitter,

and xcsgasxvbzx.

CONTENTS

AUTHOR'S NOTE

In this book I attempt to share some of what I have learned since the idea of a book about flight came into my head. I am an enthusiastic layman, only too well aware that in tackling the subject I am encroaching on areas both way above my pay grade and the specialist subjects of people far more qualified than me to write about them. I have come to some of these subjects completely cold, like an eager first-year student primed for knowledge. Inevitably, what I have learned is far outweighed by what I have still to learn. I am indebted to the many people who have devoted their lives to the subjects covered in this book, and whose work has inspired and educated me in its writing. I attempt to acknowledge all sources of information.

It's my hope that I've written about this fascinating subject in a way that will place the reader in the 'sweet spot' – somewhere between reeling from too much information and tutting at the banality of it all. If I fail, I can only apologise to everyone. Please try to bear in mind that I meant well.

All mistakes are, of course, my own.

INTRODUCTION

[illegible]

INTRODUCTION

A January Friday. Cold, grey, bleak. A day for buttered malt loaf and window staring.

It's a typical garden scene. Grass, borders, fences, shed. Bird feeders, primed. My eye is caught by a quiver in the hazel. Quiver becomes rustle, rustle becomes blue tit, hopping up to an outer branch, checking for danger. Alert, always alert. It flies up to the feeder, flicks its tail, turns, pecks, turns again, pecks again. Another flick, back to the perch, check for danger, and off it flies, over the fence and away. Whizz, whirr, flurry. Ten seconds later and I would have missed it.

It's an everyday occurrence, repeated hundreds, thousands, millions of times daily. Blue tit's gonna blue tit, and I'll always watch. They're active birds, constantly on the move. Agile, acrobatic, endearing – all three at once when they're hanging upside down from a bird feeder. But from this brief encounter one thing stands out.

It flies.

It. Flies.

It's a thing so normal, so entirely taken for granted, that we forget how extraordinary it is. Our planet has gravity strong enough to draw objects of mass towards its centre – to defy that in pursuit of an aerial life seems needlessly perverse. It's a notoriously difficult thing to do.

The trick, as Douglas Adams so neatly put it, is to throw yourself at the ground and miss.

Humans are terrible at it. We manage the first bit well enough, but it's the missing that causes problems. Gravity's a bugger that way. But this hasn't stopped us trying. Human history is littered with the bodies of those whose ambition to fly has been brought crashing down to earth – literally and metaphorically – by harsh reality. For centuries the best we could hope for was a kind of mitigated plummet.

But while our bodies are ill suited to staying up in the air, our brains are good at problem-solving. A comparatively short time after our emergence as a species we came to understand that the ways of the air, while complex and invisible, are quantifiable, and that we might be able to exploit them to our advantage. And, crucially, we got good enough at technology to make machines that could help us realise our dreams.

And so we managed to haul our stubbornly unaerodynamic bodies into the air, and – and this is the important bit – keep them there. We've been doing it for well over a hundred years now, and we're pretty good at it. With the help of technology, we can swoop, glide, hover, dive, climb, soar, float, drift and – if so inclined – slip the surly bonds of Earth and touch the face of God.* We can fly

* 'I have slipped the surly bonds of Earth . . . and touched the face of God', from 'High Flight' by John Gillespie Magee Jr.

high and low, fast and slow, round the world and back again. Thanks to flying machines, our world has changed.

But still an inconvenient truth remains: without artificial aids we are drawn by the laws of physics back to the ground. Even though we have used our oversized brains to devise ways round this fundamental inadequacy, put us in a flying race against a mosquito or a beetle or even that reluctant last-resort flyer the red-legged partridge, and there will only ever be one winner. When it comes to telling the story of flight in the animal kingdom, we must – sorry, Icarus – allow others to take centre stage.

While many species, for various reasons, remain earthbound, plenty of others happily spend between 1 and 99 per cent of their time in the invisible medium we call 'the air'. Unknowns being what they are, it's difficult to deal with exact numbers, but of the more than 1.5 million described animal species on this planet, the vast majority have the gift of flight.

I suspect we don't think about this enough. What would happen if we spent more time looking up, drinking in the spectacle of an animal defying gravity, and thinking about how this came to be, what it means, and what a wondrous thing it is to fly? At the very least, it would divert us for a few seconds. And we might imagine ourselves in their place, might strive to abandon for once our narrow, limited view of the world, see it from a different perspective. An exercise in empathy, an attempt to find within ourselves the capacity to be, however temporarily, something else.

The ability to fly seems miraculous to those not endowed with it, but a miracle is something that breaks the laws of nature – while flight is indeed awe-inspiring and envy-inducing, it sticks to the rules. The principle is in fact remarkably simple: four forces – lift, thrust, weight and drag* – combine, to the advantage of the aspiring flyer. Produce enough lift and thrust to counteract the weight and drag, and off you go. But the more you explore the subject, the more complex and involved it all gets, and the more accurate that word 'miracle' seems.

None of this worries a flying animal in the least. A pied wagtail, bouncing merrily over my head on West Norwood High Street with a cheery *chizzick!*, isn't thinking of wing loading or aspect ratios or any of the other concepts that govern its capacity for merry bouncing. It just does it, in the same way that I, when catching a cricket ball, am not doing differential calculus to work out the ball's trajectory. I merely catch it.†

The pied wagtail's insouciance is replicated manyfold in its avian peers. From the ridiculous agility of a hummingbird to the powerful grace of an eagle, the freedom that flight represents is most readily observable in birds. Of the planet's other flyers, insects are small, bats fly at night, and pterosaurs are extinct – but birds are right there, flaunting their ability for all to see.

There are many reasons to love birds – their featheriness, their behaviour, the fact that they're dinosaurs. But at the heart of it is flight. As a child, I loved birds without understanding why; as an adult, I realised that flight played a large role in my obsession. This was in direct opposition to my own relationship with being in the air. For a long time flying fell into the same category as rugby: I was a keen observer, but an extremely unwilling participant. While it

* Loosely: lift – upward; weight – downward; thrust – forward; drag – backward.

† Or, more often these days, drop it.

gave me pleasure to watch a great tit flitting from tree to feeder or swallows hawking midges from the surface of the local pond, my own forays into the air were fraught with nerves. It took me many years to embrace flying as an acceptable form of travel. Only gradually did I understand my own fear, realising that it wasn't flying that scared me. On the contrary, I was captivated by the feeling of being suspended high in the air, seeing the world laid out in miniature below me. It brought a sense of freedom that transcended the inconvenience and discomfort associated with being trapped in a thin metal tube. I wanted little more than to dispense with the material of the aeroplane and engage completely with the flyingness. It wasn't being in the air per se that gave me the heebie-jeebies. No, what I was afraid of was the idea of transitioning precipitously from flying to not flying.

In other words, I was terrified of crashing. And I didn't anticipate the pre-crash plummet too keenly either.

Over time I learned to take a more rational approach to this phobia, and in doing so I began to think about flight in all its manifestations, and to delve into what it means to flyers and non-flyers alike. Once you start thinking about the extraordinariness of flight, it's difficult to stop. Difficult, too, to quell the urge to grab passers-by, point at the carrion crow rowing gamely through the air, and yell, 'Look! Look! It's flying!'

Curiosity piqued by the wonders of avian flight, I was soon asking questions about all the other flyers. How do hoverflies hover? Why are bats the only flying mammals? What does a daddy-long-legs actually achieve with its frankly pathetic, gangling excuse for flight? How and why and when did all this airborne nonsense come about in the first place?

This last question, the question of 'when', necessarily involves some delving into the issue of 'geological time', a subject apparently designed to induce giddiness. Perhaps this is because true understanding would demand we acknowledge how utterly tiny we are, and I don't think humans are very good at that. The idea of ten years is manageable – we can remember what we were doing ten years ago and can at least have a stab at predicting what we might be doing in ten years' time. A hundred years, while more than most people's lifetimes, is still touchable – a matter of a few generations. A thousand is conceivable – think William the Conqueror and so forth – but requires some imagination. Ten thousand is familiar as a number because it's how long agriculture's been around, but try to touch it and we find ourselves groping in the dark. And as for a hundred thousand years – nope. A million? That's one hundred spans of agriculture; three and a bit times longer than *Homo sapiens* has been around; or, if you find it easier, ten nopes.

And yet a million years, in geological time, is relatively small beer. Take the lifespan of this planet. If I were to represent this time frame – four and a half billion years, give or take – using a scale of a million years per page, this book would be 4,550 pages long. The first 500 pages, it has to be admitted, would lack interest for the student of life,* although the true enthusiast might gain a certain amount of pleasure from contemplating the texture of the empty pages. But gradually marks would appear, each one representing a life form. Small marks, indecipherable at first, eventually coalescing into letters, words, sentences, paragraphs. And soon each

* There was of course a huge amount of geological activity going on, but that is not our concern.

page would be seething. At about page 4,200, the first animal flies. Humans turn up in the last paragraph of the last page.

A gimmick it might be, and I would certainly fail to persuade my publishers – obsessed as they are with such trivial concerns as the cost of paper and the practicalities of producing a four-and-a-half-thousand-page book – that the symbolism would be an important part of the storytelling. But it would be. That physical representation might help us overcome the vertigo brought on by the idea of huge, unimaginable tracts of emptiness.

Add to this the question of geography. It's easy to think about the world before we came along as existing not just some*when* else, but some*where* else too. Our brains, faced with the fatal combination of impossibly large numbers and the accompanying realisation of our own insignificance, melt a bit round the edges and refuse to cooperate. On coming across a fossil at the beach, we can cognitively understand that this is the preserved body of something that lived a very long time ago, but just as freaky, in my view, is the idea that it was *here*. Exactly here, where I'm standing right now.* This all happened in our world, and it is amazing.

Those fossils, for all their abundance, give us only the tiniest slice of information about how life grew on this planet. Although there are frustrating gaps all over the place, the existing record has enabled Very Clever People with infinite patience and a taste for puzzles to piece together what happened. Or might have happened. Or 'could plausibly have happened and let's leave it in the box marked "possible" until something better comes along'. Because

* One does of course have to allow for the shifting of landmasses that has occurred over those periods, but I hope you're with me in the basic concept.

that's how it tends to work. Scientists construct hypotheses based on the best available evidence. The responsible ones generally shroud these hypotheses with caveats: 'we think', 'it seems', 'one possibility is . . .'. And then another piece of the puzzle is found and it either confirms what they thought (hurrah!) or contradicts it (boo!) or makes the whole picture even more confusing (huh?). In some cases the body of evidence is strong enough to be considered fact, or near as dammit. In others it is frustratingly scant, and intelligent hypothesising is the order of the day. It is on this hypothesising that much of our understanding of evolution, and particularly the evolution of flight, is based.

Take yourself back a few years. Four billion should do it. At the bottom of the ocean, in the bubbling and roiling warm water of a hydrothermal vent, a microbe lives. That microbe is known as LUCA – Last Universal Common Ancestor. LUCA is not the first thing to live, but it is our most recent ancestor – where 'our' means 'all life on Earth's'. Strange though it might be to think of a piece of seaweed as our cousin – or a moss or an earthworm or even a trilobite (RIP) – that is the long and the short of it. Impossibly distant cousins, but cousins nevertheless. All life on Earth is related. We forget that all too easily.

From LUCA came other things. Then others, and others. And on it went, in agonisingly slow increments, developing and diversifying and changing. At any point, life could have taken a different route, diverted by different conditions, different circumstances, pure chance. Replay the story a thousand times, a million, a billion, and each time the outcome would be different. In those many possible iterations of this planet's history, would there be flight? Would something at some point, somewhere, somehow take to the air? You'd like to think so. But the thing to remember is this. Flying is hard. Some creatures do it so well we're fooled into thinking it's easy. But for all of them – even those so well adapted to the challenge that they can stay airborne for what we consider serious amounts of time – what it does require is constant work. Sustained, coordinated and concentrated work. Without that, a flying thing turns into a falling thing, locking horns with gravity and losing.

Flight has evolved four times. That is to say, powered flight – the act of propelling yourself into the air and staying there. Strictly speaking, flight is a continuum. Anything that spends its time at least partially airborne belongs somewhere on it. The main thrust of this book is at the powered end of that continuum. Gliding, parachuting, falling with style – they're all impressive in their own way; but flying – that's something else.

Those four evolutions of flight occurred (as far as we know) in four distinct groups of animals: insects, pterosaurs, birds and bats. And in each of those groups the ability enabled success. The diversity of species adopting it is testament to its usefulness. Fruit flies and pterosaurs are wildly different animals, but they are united in their ability to throw themselves at the ground and miss. An exclusive club of miraculous exploits.

The club's members outnumber the non-members. This is largely thanks to the extraordinary diversity of insects. How many species are there? Add to the 1.5 million described species all the others we either haven't found or named yet, and estimates range from 2 to 30 million, with 5 million considered by many a reasonable guess. Throw in nearly 11,000 bird species and just over 1,400 bats, as well as all the extinct species in every category (including at least 200 pterosaurs), and we quickly reach numbers every bit as bewildering as those associated with the enormity of geological time. And most of them fly.

From those millions of animals I have chosen fourteen. The choice is, to an extent, guided by personal preference. I've already mentioned my love of birds, but the inclusion of six of them in the fourteen isn't just favouritism – they exhibit more variety in the way they fly than any other group. Also abundant in the list are insects, the first animals to take to the air. The six I've chosen represent not just their evolutionary history, but the various bodily adaptations that have enabled their extraordinary success. Completing the list are pterosaurs – standing alone as the only completely extinct flying group – and bats, our closest flying relatives and the only flying mammals. Between them, these fourteen representatives chart the chronological and evolutionary history of flight, represent its various uses, and illustrate the different mechanisms and body plans by which it is achieved. Some – hummingbirds, dragonflies and bats, for example – do it with extraordinary agility; others – albatrosses, bees and geese, say – have particular specialities; and some – mayflies, beetles and particularly the bumbling *Archaeopteryx* – represent the 'inexpert but perfectly useful for our current needs, thanks very much' corner of the clubhouse. There's space, too, to consider what happens on the rare occasions that flying animals relinquish their superpower and join the rest of us back on the ground.

We start, as you might expect, at the beginning.

1

THE MAYFLY

Take yourself, if you can, to a stretch of clean, fresh, flowing water. It will be a spring day. There will be, if you get it right, warm sunshine. And the place will have an air of tranquillity, such as might inspire feelings of well-being in the human soul. The trickling of water, the swish of reeds, perhaps even the gentle susurration of wind in willow branches. Birdsong will probably fill the air. It often does. Everything combines to give the scene an indefinably timeless quality. You'll sit awhile, allowing time to ooze by. You will have left your phone at home (because you're sensible), so all you have to do is relax, observe and wait.

It might be a flickering out of the corner of your eye, a fracture in the continuum. Or perhaps there'll be a disturbance on the water's surface, a small splash. And then another. And over there, another. And before you know it the air is full of activity, whirring specks of fluff caught in the sunlight, a seething cloud of them, so profuse that even when you try to follow the flight of an individual it is quickly absorbed into the general melee. And meanwhile the trout pop to the surface and gorge themselves, relishing the sudden appearance of a free buffet.

The mayflies are emerging.

As you sit by that stretch of river, it's worth taking a few seconds to contemplate the idea that time travel does exist. The idyllic scene described above – or something like it – has been part of the planet's story for nearly 300 million years. Of all the flying insects in the world today, mayflies are often considered the most primitive, providing us with a link that runs nearly all the way back to the origins of flight.

Nearly.

Because they weren't the first. We don't know what was, or when it did it, or how or why it went about it, but at some point, somewhere, somehow, something took to the air and embarked on our planet's first powered flight.

We do know it was an insect. And we can be fairly sure it was sometime between, say, 400 and 325 million years ago. This isn't a narrow window, even by the standards of geological time. For that, as with many things, we can blame the fossil record.

The trouble with the fossil record is that there's simply not enough of it. Imagine watching a film, but the projectionist shows you only a quarter of a second every eight minutes. That is the fossil record.

How convenient it would be, how satisfying, to be able to slot it all together, to dispense with 'probably', 'may have', 'some think that', and all the other weasel words that necessarily accompany any discussion of our world as it was for the millions of years before we turned up. But, on the other hand, then where would we be? We would know everything there is to know about life on Earth, and that would be a different kind of awful. Having all the answers isn't necessarily all it's cracked up to be. Curious ignorance can be healthy.

Not everything that lived is preserved. Very far from it. And what we do have is sometimes the scantest trace, offering tantalising clues as to what kind of creature they might represent. And the more fragmentary the clues, the more vigorous the debate about their meaning.

Take, for example, *Rhyniognatha hirsti*,* a fossil from the Rhynie chert,† near Aberdeen in Scotland. It's about 402 million years old and consists of parts of a creature that might – just might – be the world's first flying insect.

* Many people, including me, are understandably nonplussed by long scientific names. But they're useful – you might say essential – in avoiding confusion about the name of a species, and usually have logical origins, even if deciphering them does require a little knowledge of Latin or Greek. In this case, the name derives from *Rhynie* (the place where it was found), and *gnath* (meaning 'jaw'). Hirst was one of the authors of the original report describing the fossil in 1926.

† I had to look up 'chert'. Maybe you did too. To save you the bother: 'a siliceous rock of cryptocrystalline silica occurring as bands or concretions in sedimentary rock, e.g. limestone' (Chambers).

The word 'might' is doing some heavy lifting here, because what exactly *Rhyniognatha* was remains a matter of some debate. A layperson looking at photographs of it, an indistinct agglomeration of brown shapes on a yellow background, might be reminded of a Rorschach test. There is certainly nothing even remotely resembling a wing in there. Those shapes are in fact head parts, and when those are all you have to work with, conjecture and deduction are a necessary part of the process. Any advanced skill can seem like magic to the uninitiated, and so it is with fossil identification, where the ability to build a convincing picture of a creature from the tiniest sample is crucial. 'How do you know?' we ask, stupefied. The answer is usually, 'Well, we've spent a very long time thinking about it.'

Discovered in 1919, *Rhyniognatha* was first described a few years later as a springtail-like creature, and so the situation remained until 2004, when American researchers Michael Engel and David Grimaldi looked closely at the mouth parts and concluded that their shape, size and general demeanour were hauntingly similar to those of modern mayflies.[1] As technology improves, so does our capacity for analysis. A 2017 study by paleoentomologists Carolin and Joachim T. Haug, using advanced microscope technology, cast doubt on the mayfly interpretation – indeed, the authors suggested that *Rhyniognatha* might be some kind of centipede.[2] An arthropod, then, but not an insect. It remains to be seen whether this is the final word, and the case for both interpretations is hotly argued.

There is equal uncertainty surrounding the fossil fragments of an insect from about 385 million years ago found in New York State. Bits of cuticle, a compound eye. What exactly was it? Something insecty, wingless, silverfish-like. Beyond that, it's difficult to tell.

The same can't be said for the remains of *Delitzschala bitterfeldensis*.*[3] This one is the real deal, a creamy white imprint on dark shale, the outline of its wings clearly preserved in the rock. You can see not just their shape but the fine tracery of the veins and irregular little white spots decorating the wing. My imagination wants to turn these spots into glowing orange or iridescent purple – something to bring the animal even more vividly to life.

It's about 325 million years old, this fossil. Wingspan of a couple of centimetres. The wings look both familiar and advanced – you wouldn't be surprised to see such a creature today. The order it belonged to, the Palaeodictyoptera,† is extinct. And it's not fully established how they are related to modern insects. But the shape and vein pattern of the wings – elegantly rounded ellipses held upright – are strongly reminiscent of a mayfly's. Is it fanciful to close your eyes and imagine yourself sitting in an aquatic environment, watching it flutter weakly among the ferns? Yes, it is. But it's also irresistible.

The problem is this. We have no idea what led to *Delitzschala bitterfeldensis*. You might expect there to be clues in the fossil record as to how such an advanced wing came about. But between the silverfishy fragments and *Delitzschala* we have, to all intents and purposes, nothing. Not just a lack of winged insects, but barely any insects at all.

* In this case the name is derived from the two places in Germany, north of Leipzig, nearest where the fossil was found.

† The word pieces together the stems of Greek words: *palaeo* ('ancient'), *dictyo* ('network') and *pteron* ('wing').

Welcome to the Hexapoda gap – the period from 385 to 325 million years ago, spanning the end of the Devonian period and the beginning of the Carboniferous.

The Hexapoda gap is like a tunnel. At one end, the mid-Devonian.* At this point Gondwana, the supercontinent dominating the south of the planet, has been around for about 160 million years. Atmospheric levels of oxygen are holding steady at about 15 per cent, after dipping down below 13 in the Silurian. Vascular plants are spreading and beginning to form forests. Arthropods are wriggling and crawling their way across the ground. Life, for so long the preserve of the water, has established itself on land.

We emerge from the tunnel 60 million years later, in the early Carboniferous. Gondwana is in the process of merging with Laurussia to form the even larger supercontinent Pangaea. Oxygen levels have risen to above 20 per cent. The plants that were groping their way towards abundance and diversity at the beginning of the tunnel have gone whoosh, spreading and diversifying like nobody's business. Flying insects have evolved to the extent that we have *Delitzschala bitterfeldensis,* and possibly plenty of others both like and unlike it – proof that the next barrier, between land and air, has been breached. But in the tunnel, there's 60 million years of nothing much. Whatever insect life was active in that time was either not preserved or hasn't yet been found.

Insects being, in the main, small and delicate, the conditions required for their preservation in the fossil record aren't encountered that often. The hard grey rock of the Rhynie chert is an outlier. Here volcanic springs brought forth water rich with

* To be precise, the Givetian stage.

silica, immersing and petrifying everything in its path. A key factor was that the preservation was almost immediate. As a result, the many plants of the Rhynie chert are preserved in extraordinary detail, and along with them a decent chunk of fungi, lichens, algae, cyanobacteria and arthropods, this last category including the aforementioned *Rhyniognatha hirsti*.[4] What we need is a Rhynie chert from the early Carboniferous, yielding just one insect fossil that might answer questions about the development of the insect wing. It's not too much to ask, surely?

Lacking such information, we have, as yet, no clues as to whether *Delitzschala*'s relatively advanced wing was the result of slow and steady evolution throughout the Hexapoda gap or a sudden radiation near the end. What we do know is that it, and others like it, flew, even if exactly how they first did it remains a subject for conjecture.

Evolution being what it is, the sudden appearance of complete, complex wings, where there were no wings before, is not an option. Evolution isn't interested in what might happen at some unspecified point in the future. It works only with what it has to hand. If an adaptation offers a benefit, that adaptation is favoured. This leads to the old question, so often asked by those dubious about the credentials of evolution by means of natural selection: what use is half a wing?

Plenty, as it happens. The point is that an evolutionary adaptation merely has to be an improvement on what was there before, offering a better chance of survival, however tiny. In the same way that eyes didn't start as fully developed eyes but merely organs that could distinguish between light and dark, the complex flapping wing started as something much simpler.

But what? And why? And how? Such insects as existed 400 million years ago had barely got used to being out of the water and on the land. What factors might have set the ball rolling on the gradual development of wings?

While many explanations have been proposed, they boil down to two basic ideas.[5]

The first – the paranotal lobe hypothesis – has little veined lobes sprouting on the front section of the insect's thorax. These lobes might have had some other function, possibly acting as solar panels to help the animal warm up. The added advantage was that they helped stabilise it as it dropped from the top of a plant to the ground, offering at the very least a softer landing, and possibly the ability to glide a short distance. Those with better stabilisers were more likely to survive, so they evolved to become even better, and gradually turned into something that conferred a more tangible aerodynamic advantage. At this stage this would merely have taken the form of being able to travel a little bit further before hitting the ground – from plummet to glide, however short. As they developed, the gliding distances became gradually longer. From stabilisation to short-distance gliding, and then from there – with the addition of an articulation where wing joined to body – to actual 'Look at me, Ma!' flapping flight, all in the relatively short space of a few million years.

Counter to this proposal is the – to this layperson equally plausible – gill hypothesis, which points to the gills on the larvae of things like the modern mayfly. They use them for breathing, but they also resemble tiny wings, and while mayfly nymphs swim using their tails, if someone said to you, 'Those gills could easily, over time, grow big and strong enough to propel something', you'd have a hard time denying it. Add to this mix the convenient fact that the gills have veining patterns tantalisingly similar to those on insect wings, and it's easy to see the appeal of this hypothesis.

Lobes or gills? For those of us naturally inclined towards compromise, a recent proposition is even more attractive: namely that the truth is a hybrid of the two hypotheses. Given that each has wings originating from different parts of the body, this might seem counterintuitive, but the 'dual origin' hypothesis – that insects' wings were the result of a gradual merger between paranotal lobes and gills – is, to lapse briefly into racing parlance, coming up on the rails, full of running.

Whatever its origins, the wing quickly became indispensable. And with good reason. A well-made wing is a clever thing. It collaborates with the air to keep a body off the ground. And it does it (and this is the clever bit) merely by existing. A well-made wing is, you might even say, a wonder of nature.

If you were asked to design a wing from scratch, instinct might guide you towards a workable prototype. Biggish, lightish, flattish. A needle is not a wing. Nor is a brick. A piece of A4 paper, on the other hand, is at least a decent starting point, because it has a large surface area for its weight, and that larger surface area produces more lift. But a piece of paper is weak, so you might want to make it from something stronger, or to bolster it with strategically placed stiffeners. But they add weight, which means you need more lift . . .

And so the long day wears on.

Combining lightness with strength is one of the fundamental conundrums facing all would-be flyers. The solution evolved by insects is to make the wings out of the same stuff that is predominant in their exoskeletons: chitin. This extraordinary material – light, sturdy and waterproof – is perfectly suited to building a wing, as long as it can be made thin enough. But a single sheet – like our piece of A4 but much thinner – is too weak. So the wing is made of two layers, tightly compressed and threaded through with a network of veins. The clever thing about this is that the veins not only strengthen the structure but bring blood ('haemolymph', strictly speaking) to the extremities. But with that strength comes added weight – they are heavier than the surrounding chitin – so the proportion of membrane to vein is important. Too few veins, and the wing will be too weak; too many, and it will be too heavy.

Looking at an insect's wing with the naked eye, it appears flat, but of course veins aren't two-dimensional, and once you get them under a microscope you enter a world of plains and valleys, minuscule hairs and veining networks of fascinating intricacy. While marvelling at these details, you also see something of their structural strength and can appreciate the possibility that they might be strong enough to propel their owner upwards.

But zoom out and once again they look impossibly fragile, the thinness of the cuticle and the filigree tracework of the veining giving the impression that they'd be vulnerable to the merest puff of wind, liable to fall apart at the merest misplaced touch. With such things insects took to the air.

So, we have a wing. It is aerodynamically efficient, and strong enough to withstand any buffeting or wear and tear it might receive in the insect's short life. Now you need to find a way to move it. A static wing is OK for gliding, if it's big enough. But to get off the ground, and then to move actively, you need a form of propulsion.

Because our insect wing contains no muscle, propulsion must come from the body. The most straightforward solution is to attach the flight muscles directly to the base of the wing with some sort of simple joint. This enables what on the face of it seems a fairly basic and unsophisticated form of wing manipulation.

That – more or less – is how those first flapping wings evolved. Muscle pulls up; wing moves down. But muscles are capable of just one kind of movement – they shorten. If you want to move something in two directions – down *and* up, say – you need two sets of muscles. One to move it one way; another to move it the other. This pleasingly simple arrangement worked for *Delitzschala bitterfeldensis*. And it still works well enough for mayflies, which emerged not long after *Delitzschala bitterfeldensis* graced the airways. While those very first flying insects are long gone, our modern mayflies, built to a similar plan, give us some idea of what they might have looked like. They have long tails, flight muscles attached directly to the wings and, most importantly, wings that don't fold along their abdomen. It's this last attribute that places them in the group Palaeoptera,* a fairly loose gathering of unrelated insects with similar characteristics (what is known as a 'wastebasket taxon') in which can also be found the dragonflies.

It's one thing knowing how something could have happened; still another to understand why. In making its way from water to land, life had already overcome a significant barrier. The modifications required for transition from one environment to another are considerable. A creature adapted for life underwater will struggle on land, and it's a similar story getting from land to air. Different

* 'Old wings' in Greek. Neoptera are 'new wings'.

skill sets are required, different abilities, but mostly a different anatomy. And for all these modifications there needs to be an incentive. Gravity is powerful, and you'd need a good reason to defy it.

Not being eaten is the most immediate incentive. If you can't outrun danger, a short glide might just do it – and even better if you can flap your way to freedom, casting a backwards glance to enjoy the baleful glare of your disgruntled attacker.

There are also less immediate advantages. If you can fly you can travel further to find your own food, and you can get there faster than your leaden-footed earthbound rivals. Imagine you're a Devonian insect feeding on spores at the tip of one of the tall plants that seem to be all the rage these days. There's another plant over there, just too far away for a jump. How much time and energy you'll save if you can fly to it, without having to go through the tedious business of climbing down the stalk to the bottom, along the ground and then up the next stalk.

Looking to the medium and long term, flight enables you to disperse. Maybe this dispersal is, in the first instance, short range. This is useful in the event of your local habitat being, for example, devastated by fire. Then, over time, the range of dispersal radiates, enabled by the superior ease and speed of travelling in the air versus the slog of land travel. This is a good thing for the growth and diversification of the gene pool.

Having wings has an added benefit, not directly related to flight. If you're a cold-blooded animal, a quick-warm system is immensely useful, enabling you to get going more quickly in the morning. How useful to have a couple of solar panels strapped to your body. And once they've grown to a certain size, you can start to decorate them with bright colours and eye-catching patterns, all the better to attract a mate or warn a predator that you're poisonous and therefore not suitable snacking material.

Such benefits, accrued in tiny increments, were strong incentive for evolution to favour the development of wings in some species. But there were costs associated with it all. Flying is expensive, energy-wise, and the muscles you need to power yourself into the air are not light. What's more, once you've grown your wings, they present a problem for those times when you're not using them – great for flying, awkward for hiding in small spaces.

None of the problems associated with flight are insoluble, but the margins are small, and for many animals it's just not worth the candle. Perhaps they're too big, too heavy, not fast enough. Perhaps the difficulty of making the change outweighs the benefits it brings. Or perhaps they're successful enough doing what they're doing. But for some, flight offered an alternative way of life, and they embraced it. Insects with the ability to fly prospered, passing on their advantageous traits to their offspring, and before long (in geological terms) wings became standard equipment.

The mayfly's story is a romantic one. The insect that lives for a day. It serves as a useful prompt for us to consider the brevity of our own existence, to remind us to seize the moment, to live life to the full, not to waste time on fripperies. They are the embodiment of the ephemeral, a fact reflected in their scientific name: Ephemeroptera – 'short-lived wings'.

The only thing is, it's not entirely true.

We take as our default mayfly, from the approximately 3,000 species worldwide, the one known as the green drake. *Ephemera danica,* to give its scientific name.

The adult, sure enough, is short-lived – anything from a few hours to a few days. But before they reach that final flourishing, their nymphs have spent up to two years doing the donkey work, sitting on the riverbed, doggedly hoovering up tiny plants and algae, and going through up to twenty stages of growth before emerging to complete the cycle. That we focus on the glamorous adult stage is entirely understandable, but spare a thought for the hard-working nymph.

When the time comes, as it must, up floats the nymph to the water's surface. Some, resistant to change, expend valuable energy struggling back down to the riverbed, where everything is comfortable and familiar, rather than embracing the inevitable. The ones that save energy, submitting to whatever may happen and allowing themselves to float to the surface, are better equipped to deal with what comes next. Given that what comes next might be a hungry trout, speed is of the essence. Once on the water's surface, the nymph shucks off its restricting exoskeleton and emerges as a flying insect. Incidentally, this fact alone – so commonplace in the insect world, so abundantly practised, so pivotal to their success – remains a thing, at least to my mind, of extraordinary wonder.

The insect that emerges is an adult rather than a larva, but it is not the full adult of folkloric fame. It's the intermediate subimago – known to fly fishers as the 'dun' – and it is unique. Because in all the astonishing variety and diversity of insect life, mayflies are the only ones with two flying adult stages.[6]

Not to go on about it or anything, but I'll just say that again. The only ones. Millions of insect species on the planet, and these 3,000 or so choose a life cycle avoided by all the others.

Why? What advantage does the apparently cumbersome use of a short-lived, clumsily flying intermediate stage confer?

Good question.

Other aquatic larvae – dragonflies, for example – might crawl onto land to make the transition to adulthood. But mayflies (most of them, at least) eschew the land completely. The dun has one job: fly from here – the water's surface – to there – the river's edge, where the moult to full adulthood can take place. And do it sharp-ish, because that hungry trout's getting closer.

The dun – drab in colour, hence the name – differs from the adult ('spinner', in fly-fishing parlance) in one key area. Its wings are covered with microscopic downy hairs called microtrichia.[7] It's thought that they help slough off the water that would disable the adult's smooth, hairless wings. This is an important adaptation, helping bridge the abrupt transition from life aquatic to life aerial. Its brief time on the water's surface is fraught enough with peril without worrying about drowning. Despite this, many duns perish at this stage.

The dun is able to fly almost immediately. Its weak, fluttering flight takes it to a waterside perch, where it can undergo the final stage in its remarkable life cycle.

A resting mayfly seems to sit on its belly, the legs providing some lateral stability, the hind segments of the abdomen curving gently into the air. The three spikes of the tail (cerci) stick out like the untameable hairs of an old man's eyebrows. The eye is drawn to the rounded triangle of the forewings, held stiff and upright away from the body – in some places mayflies are known as 'upwings' – and it's only if you get a chance to look more closely that you might see the less developed hindwings lurking in their lee.

The thing about mayfly wings is that they're not very good. The other thing about them is that this is absolutely fine. They don't need to be the best; they just need to do their job. A mayfly has no need to hunt. It has no mouth parts, living instead on energy stored while in the larval stage. It has no need to migrate. Gliding, hovering, soaring and all the other activities potentially available to winged creatures are of no interest to it. All it wants is to get into the air for long enough to find a mate and breed.

The males fly up in a swarm, using their precious reserves in the energy-intensive activity of flight. It's a stumbling, fluttering flight, their abdomens and cerci hanging down below them in ungainly fashion. Their flapping is slow enough for the wing movements to be visible to the naked eye. But what we can't see is what happens to the air around the wings.

An insect's wingbeat is divided into two half-strokes, conveniently labelled 'downstroke' and 'upstroke'. The downstroke is executed more slowly, and at a deep angle, to produce thrust as well as lift; then the wing is flipped over so the leading edge is facing backwards for the shallower, faster upstroke. Repeat multiple times per second.

As well as pushing down on the air to lift the insect up, that angled downstroke also produces what is known as a leading-edge vortex (LEV* for short). The LEV is variously described as 'tornado-like', 'spiralling' or 'swirling'. Whichever description you

* Gosh, well, what can I say? It truly is an honour.

favour, its effect is to increase the speed of the air above the wing. With faster air speed comes lower pressure, and therefore increased lift. Even at the relatively slow wingbeat speeds produced by a mayfly, the combination of these forces is enough to propel them into the air. This is the method favoured by most insects.

Once the male mayfly has reached what he deems a suitable height, he parachutes down, no doubt hoping to catch the eye with the elegance of his trajectory. Fly up, float down, repeat – an elegant mid-air dance. The females fly into the swarm. A lucky male will find a female, grab them with their legs, and mate in mid-air – ill-adapted to the land, this is the most convenient place for their union. The unlucky ones, having escaped the perils on the water's surface and made it this far, will be taken by sand martins or swallows, agile birds well up to the task of outwitting and out-manoeuvring a poor flyer distracted by the urge to mate.

But the mayflies have emerged en masse for a reason. They're not alone in using the idea of predator swarming to ensure the survival of the next generation – make loads of eggs, so that when they hatch, even when the local predators have had their fill, there will still be plenty of survivors.

Copulation complete, the females float down to the water's surface, dip their tails into the water to lay their eggs – as many as 8,300 of them – and die. Their life cycle starts and finishes in water, and only the briefest part of it is spent in the air. But no matter how weakly they do it, the flying stage is crucial.

From tentative beginnings – the development of rudimentary wings, the first explorations of the air – insects made their way off the ground. For the early adopters, the innovation of flight – one of the great advances life on Earth had yet seen – enabled them to colonise the air with spectacular success. Like the Grand Old Duke of York's men, when insects were up they were up – and for them there was no going down. They diversified and spread enormously in the late Devonian, and before long they ruled the air.

It would be theirs for the next 120 million years.

2

THE DRAGONFLY

Here is something that might have happened.

It is a few years after the time of *Delitzschala bitterfeldensis*. By 'a few' I mean 25 million or so. That is to say, 300 million years ago, or page 4,250 of our putative *Bumper Book of Life*. For those who prefer a more official way of measuring, it is the late Pennsylvanian stage of the Carboniferous. We are in what will one day become the Auvergne in central France. A small insect is whizzing around, enjoying the freedom of flight. It's not the only insect around. Far from it. In fact, the air is positively abuzz. But this is the one we're going to focus on, because of what is about to happen to it.

We don't know what species it is. Let's call it a whizzfly. The whizzfly has recently evolved from a wingless lineage, and – were it capable of such things as conscious thought or emotions – would be feeling pretty smug about its ability to fly compared with the earthbound crawlings of its ancestors. Life is good. And it's even better in the air. *Whizz whizz whizz.* This whizzfly is particularly enjoying – and if we're going to invent a 300-million-year-old insect just for fun, we might as well anthropomorphise the hell out of it – the feeling it gets when it does those little darting turns that seem to impress the female whizzflies so much.

But just as it is relishing these hypothetical thoughts, it is, without so much as a by-your-leave, eaten.

RIP whizzfly. We barely knew you.

This kind of thing happens all the time, of course. But it's of interest to us, not just because it stands as a proxy for the new-found abundance of flying insects but because of the identity of the whizzfly's eater, which is – as we know from fossils first found at Commentry in the Auvergne in the 1880s – the largest flying thing yet seen on the planet. It is still today the largest insect to have lived, an ancestor of the creatures we now call dragonflies.

We have given it the name 'griffinfly'.*

By contrast with the mayflies that were beginning to take hold at the same time, and despite their similar appearance, griffinflies didn't use their dainty wings to flutter up, gossamer-like, in search of a mate; they used them to outspeed and outmanoeuvre their prey. Griffinflies – like their dragonfly descendants today – were expert hunters.

The largest griffinflies had wingspans of about 70 cm. This is similar, for comparison, to the wingspan of that common city-dweller so reviled by so many, the feral pigeon. But while this might encourage headline writers to lead with 'DRAGONFLY THE SIZE OF

* Or 'griffenfly', according to some.

A PIGEON', that would be more than a little misleading – insects and birds have wings in common, but that's about it. Their body shapes are not comparable.

Still, if you're struggling to size a pigeon in your imagination – and not to give anyone nightmares or anything – 70 cm is about twice the width of your head. Try not to think of a griffinfly flapping around in the corner of your kitchen.

Incidentally, you might have read – I certainly have – about 'dragonflies with six-foot wingspans', or some such. A couple of them make a brief appearance in Michael Crichton's *Jurassic Park*. The idea of dragonflies that size is fascinating and grotesque. It also seems to be based on no evidence at all, the suggestion apparently appearing in the public consciousness sometime in the 1980s and taking hold in the same way as 'bumblebees are too heavy to fly, actually' (more of which later) in the public consciousness as one of those Fascinating Facts about Insects.

Anyway, let us, as the French have it, return to our sheep.*

Mention of the behemoth insects of the late Carboniferous leads many people to ask why we don't have such creatures nowadays.

As usual, there are varying hypotheses, each with its own adherents and degrees of certainty attached. Most widespread is the notion that it has to do with their respiratory system. It's fairly obvious to even the most inattentive observer that insects are not like us. We have our squidgy bits on the outside, draped round a skeleton and kept in check by a resilient but pliable skin. Insects have it the other way round: squidgy bits on the inside, held in by an armoured construction of chitin – the exoskeleton. This is,

* 'Revenons à nos moutons', in the original.

I think, a familiar concept. And while we have lungs that help us breathe with moderate efficiency, insect breathing is of a different order. Air is taken in through openings in the side of the body – spiracles – and then diffused around it through a network of tubes: tracheae (larger and open-ended), and tracheoles (smaller cul-de-sacs). It's an effective, highly efficient system that works particularly well up to a certain size. But, broadly speaking, the further the air has to travel, the less efficiently the system works – hence the upper limit on insect size.

But what if the air is more oxygen-rich, as it was in the Carboniferous? Back then it's reckoned that oxygen levels reached 35 per cent (they're about 21 per cent today).[1] The more oxygen in the air, the further and more efficiently it could diffuse through an insect's body, enabling it to grow larger.

There is indeed a correlation between higher levels of oxygen in the Carboniferous and oversized insects. But, inconveniently, this doesn't explain why larger insects were also prevalent well into the next period, the Permian, when oxygen levels were plummeting. While this doesn't completely knock the 'higher oxygen levels causes insect gigantism' theory into touch, it does mean there were probably other factors at play. For example, the absence of aerial vertebrate predators, such as birds, might have enabled griffinflies and their prey to grow incrementally bigger, egging each other on in a sort of body-size arms race. The niche 'largest aerial predator' was available, and griffinflies filled it.

They had the run of the place for a while, but as oxygen levels fell in the Permian, life, one can only assume, got harder. The delicate balance that enabled these creatures to become airborne in the first place was disturbed, and it would gradually have become more difficult for them to produce enough lift to get them up into the air. Not ideal for animals whose success relied on being aerial hunters. They had their moment, but all good things come to an end. About 252 million years ago came the extinction event known as 'The Great Dying'. Whatever its cause – current money is on a huge surge in carbon dioxide levels after a massive volcanic eruption in what is now Siberia[2] – it was particularly devastating for marine animals, and the largest extinction of insects the planet has seen.[3] Around that time, the fairly scant fossil record of the griffinflies fizzles out. But while they didn't survive, their place was taken soon enough by their smaller, more advanced and even nippier relatives – dragonflies and damselflies – which now enliven the air above freshwater habitats all over the world.*

Let's get the admin out of the way. The word 'dragonfly' is often used to encompass the entire order Odonata, but this in fact divides into two suborders: true dragonflies (Anisoptera, or 'dissimilar wings') and damselflies (Zygoptera, or 'similar wings'). The distinction is accurate enough – the hindwings of a true dragonfly are slightly broader than its forewings – but not always obvious, and not the first point of identification. My inexpert method is 'if it looks like a winged matchstick, it's a damselfly'. True dragonflies tend to be larger and bulkier, as well as giving an impression

* Wherever you see the words 'all over the world' in relation to the distribution of species, you can usually safely add the words 'except in Antarctica'. Feel free to apply this rule throughout this book, and especially here.

of strength in flight. And at rest they adopt different postures – damsels with their wings aligned to the body, dragons with wings sticking out to the sides.

Admin over.

Here's what dragonflies can do. They can fly fast. They can hover. They can glide. They can fly backwards. If you turn them upside down and drop them they can right themselves as near to instantaneously as makes no difference. They can jink to the left just when you think they're going straight on. They can dart to the right just when you think they're going to jink to the left. They can turn in the air as if on a sixpence – and they can do it without getting dizzy or vomiting. It might be easier, in fact, to enumerate the things they can't do. If you're looking for mastery of the air, perfectly adapted to the requirements of the flyer, look no further than the dragonflies. And all this with apparatus that has remained fundamentally the same for about 200 million years.

Compare the aerobatic prowess of these flying ninjas to the clumsiness of the mayflies. This might seem unfair – they are different animals, with different needs and expectations. But the contrast between them gives us an important staging post in the story of flight. As the only living insects with the non-folding wings and direct flight muscles common to the Palaeoptera, mayflies and dragonflies are united by their body plan. Mayflies use it in a limited way, to help with a short and specific part of their life cycle, but for dragons and damsels (and griffinflies before them), it's the means to an entire lifestyle. With it, they became among the most efficient hunters in the animal kingdom, with a success rate of around 95 per cent.[4] They're the perfect advertisement for its possibilities, and a reminder that such simplicity is not to be sniffed at.

They owe this expertise to a combination of factors, each one worthy of wonder. Take the wings, for example. It's not just their appearance, although that in itself is, to this observer at least, fascinating. The subtlety of the veining reveals itself only on close examination. Like the A-roads, B-roads and motorways of a road network, there are different grades or thicknesses of vein. The thickest – and strongest – runs along the leading edge: the costa. While insects are small enough for efficient streamlining to be less important than it is for larger flyers (for reasons we'll come to), it's still good to have strength in the bit of the wing that cuts through the air. And the leading edge has other features that at first might seem insignificant or purely decorative.

Near the tip of each wing, about two-thirds of the way along, there's a dark spot or bar. These are the pterostigmata – 'wing marks' – and they are wonderfully subtle stabilising tools, a bit like the little blobs of Blu Tack you stick to the underside of a paper plane's wings in a doomed attempt to stop it from swirling out of control and crashing to the ground. Only much better. They're a little thicker than the surrounding veins and have the effect of adding weight to that point on the wing. It seems like a small thing, an insignificant advantage, but that extra weight in the right place can increase their maximum gliding speed by up to 25 per cent.[5] Griffinflies, incidentally, lacked pterostigmata, as well as having much simpler vein patterns.

Closer to the body, at the juncture of two veins on the leading edge, is a little dip in the shape of the wing. This is the node, and it is packed with resilin. This elastic, rubbery protein occurs in many insects and is incredibly good at storing energy. Its role here is to endow the node with a little 'give', allowing the wing to flex and distort to a certain extent. The advantage this flexibility gives the otherwise rigid wing in flight is appreciable and makes up – at least in part – for the absence of muscles in the wing itself. The closer the node is to the body, the more the wing beyond it can distort, and the nimbler the dragon or damsel.

Dragonfly wings are large, as insect wings go, and, as you might expect, are therefore not the fastest. True dragonflies manage the comparatively slow speed (for insects) of around 20 beats per second, while damselflies – smaller, lighter – generally operate at about 35.[6] But what dragonfly wings do have going for them is control. A dragonfly can control each wing individually, deploy them in front and back pairs, flap them in unison or favour the left over the right. With this control they're more than a match for their prey of choice – usually other insects ranging in size from small midges to fellow Odonata not much smaller than themselves.

A simple dragonfly flight from A to B might go like this. To take off, it briefly flaps its two pairs of wings in unison at a deep angle of attack – this generates enough lift to get it off its perch, but no thrust to move it forward. But soon – within a couple of wingbeats, or about a tenth of a second – it changes pattern, flapping the hindwings slightly faster than the forewings and decreasing the angle of attack. This unsynchronised flapping moves it both up and forward. It might then switch to its default mode: both pairs flapping at roughly the same speed, but the hindwings lagging behind the forewings by about a quarter of a stroke.*[7] This works out – once

* Among other insects, grasshoppers also do this, but they lack the ability to change the synchronisation of the wings at will. They do have other advantages, though, not least leg muscles that would be the pride of any track cyclist.

you've factored in the complexities of turbulence and vortices produced by the beating of the wings – as the most efficient flapping style for most dragonflies.

If conditions are right, and it fancies saving some energy, it might have a bit of a glide, holding its wings flat and allowing the air to do the work for it. Their comparatively large wings might not be suitable for rapid beating, but they're good for gliding – a relatively rare ability in insects. What is needed for successful gliding is a wing with a high aspect ratio – the number produced by dividing wingspan by average wing chord (distance from front to back). Albatrosses are a good example of this – they have long, narrow wings with an aspect ratio of about 15; by contrast, the short, stubby wings of a sparrow yield a comparatively low aspect ratio of about 5.[8] Dragonflies' wings typically have aspect ratios between 10 and 12, which puts them firmly in the category 'can definitely glide, given the right conditions'. Damselflies, though – smaller and more delicate – are not gliders.*

The great thing about gliding is that you do nothing. Just spread your wings and float on the air. The bad thing about gliding is that, gravity being what it is, this only works for so long. If you can find pockets of rising air to give you a little boost, you can keep it going for a while, and it is obviously better than plummeting, but in general, the only way – however gradually – is down.

If the knack to flying is to aim for the ground and miss, the knack to gliding is to do so in as carefree and nonchalant a manner as you can muster. The knack to hovering, though, is of a different

* This is so nearly completely true, but one family bucks the trend – the unusually large forest giants, or helicopter damselflies.

order – keep on missing the ground while appearing to remain absolutely still.

This, it's safe to say, is difficult.

True hovering requires a very great deal of effort. What you're doing, in effect, is suspending your entire body weight in the air using only your own strength. There are varieties of hovering that involve cunning use of headwinds – this is of course a valuable skill – but I like to think that the true, purist hoverer regards such techniques as a form of cheating.

One attribute that will help guide you towards a successful hovering career is to be small. Swans are inexpert hoverers. They simply weigh far too much – even taking off is quite the palaver, albeit one that is mesmerising to watch. There are smaller birds that can hover in a way that reminds me of how I used to tread water when I belatedly learned to swim in my teens – fleetingly, panicking, and in the knowledge that this is not a state that can be maintained for more than a couple of seconds. Small birds such as finches and tits are good at this kind of hovering.* But true hovering is really the preserve of insects, whose strength is far better matched to their size.

Compared with small birds, dragonflies are hovermasters, and at least part of this is down to their wing control. A hovering dragonfly executes a downstroke with one pair of wings that it simultaneously cancels out with an upstroke with the other pair.†

* Yes, yes, I know – hummingbirds. I promise I will deal with them. How could I not? Hummingbirds – in almost every possible way – are *different*.

† Counterintuitively, this is the method – called 'counterstroking' – that damselflies use for forward flying. Different strokes (and different physics and aerodynamics) for different folks and all that.

The method sounds simple enough, if strenuous, but throw in the instabilities produced by the beating of the wings themselves, as well as the continual adjustments necessitated by air flow's irritating habit of not being predictable and constant, and it very quickly becomes a nightmare of fluid dynamics and impenetrable equations. True dragonflies are pretty good at all this, although they don't quite achieve complete stasis in their hovering and are still reliant on canny use of wind currents. But damselflies, generally even smaller and lighter than dragons, use the same technique and shallower wingbeats to even greater effect.

The extreme effort required to maintain apparent motionlessness in the air is a sign of the inherent instability of dragonflies – and indeed of all animal flyers. This, counter to what you might expect, is a good thing – it's what enables them to be so manoeuvrable. They don't have an 'autopilot' setting and their flight is constantly subject to a tweak here and a nudge there to keep them on track, but that's all part of the deal and means they can respond to impulses from their sensory system and make minute adjustments to trajectory and speed. If an aeroplane were as unstable in the air as a dragonfly, the pilot would find it almost impossible to fly, besides giving the passengers the ultimate in thrill-vomit rides. Most human-designed aircraft are made to be more or less stable. That way they can fly with minimal human intervention.* But a dragonfly wants to go over there and down there and now here and up a bit and left a bit and down a bit and stop and *look there's*

* The exceptions here are modern fighter jets, for which manoeuvrability is naturally a big advantage. Advanced computer technology enables them to be more like dragonflies than most passengers would be comfortable with.

a fly go go go I've got it, Steve, I've got it and back up round there for a bit of a rest. And it wants to do it all quickly. It also wants to do it backwards.

Backwards. That's the one that ever so slightly makes me go 'huh?'

It turns out that the technique is simplicity itself. All you do is swivel your body from the horizontal to the vertical and keep flapping. The lift that would normally make you rise is now shifted by 90 degrees – the direction of lift is relative to the flying body, not to the ground – with the result that instead of going upwards you're going backwards.

It is, as always, a bit more complicated than that. You might as well say to a human, 'To walk up the wall you just rotate your body 90 degrees and walk normally.' However, the dragonfly is well equipped to overcome the different aerodynamic challenges created by holding the body vertical and moving it backwards. Its wings beat with a deeper angle of attack and are subject to increased deformation in the struggle to overcome the unusual forces working on them.[9]

These abilities combine to make a portfolio of aerobatic excellence. But the mechanics are only part of the story. It's one thing having a body plan built to the highest specification of speed and manoeuvrability, but the engine counts for little unless it has a matching guidance system. And this is where the eyes come in.

The nature of a dragonfly's vision is different from a human being's. Our two simple eyes do very nicely for our purposes, but in a game of Animal Kingdom Ocular Top Trumps, the dragonfly's would be up there among the most coveted.

Most of a dragonfly's head is made up of two compound eyes – the largest eyes, proportional to body size, of any animal. Each one is an agglomeration of up to 30,000 facets, and each of those facets has its own lens, producing an image it sends to the brain for processing. The facets are arranged so the dragonfly has, as near as makes no difference, all-round vision. That vast range of information produces a detailed picture of the world, which is of a different order to what we call vision. In some ways the eyes might be more accurately described as motion detectors – the dragonfly's vision lacks our binocular acuity and is what we would regard as heavily pixellated, but it is much better at detecting movement. The facets don't all have the same capability – some are devoted to looking downwards, for example, and there are some that cover the area directly in front of the dragonfly in more detail.

And then there are the opsins. We have three of these light-sensitive proteins in our eyes, enabling us to see combinations of red, blue and green. In other words, along with many other mammals, we have trichromatic vision.

'I see your trichromatic vision,' says the dragonfly, 'and I raise you tenfold.' Tests on dragonflies have revealed them to have between fifteen and thirty-three opsins, depending on the species.[10] So, their visible colour range – including ultraviolet – is far greater than ours.

Add to the flashy glamour of the compound eyes the unheralded work done by the ocelli. These are simple eyes in the middle of an insect's head. They are thought to help the insect balance and stay aligned to the horizon. Without this capacity the insect would fail to stay in the air.

Bringing all this together is the brain. It processes the

information and gives instructions to the body to make coordinated movements. So far, so normal. But amazingly, dragonflies possess neurons that give it capabilities not dissimilar to those of mammals. They can focus on a small object moving against a busy background, follow it and – to a certain extent – anticipate its path.[11] This is sophisticated stuff and combines with the manoeuvrability to account for the aforementioned 95 per cent prey hit rate.

The dragonfly's mechanical proficiency, its optical acuity and the brainpower required to process the stream of information – all of this is simultaneously extremely primitive and extremely advanced. So primitive that the technology for it evolved around 300 million years ago and has survived basically unchanged since then, and so advanced that we humans, proud of our problem-solving ability, and curious and sophisticated enough to investigate and examine and analyse what dragonflies do and how they might do it, are only just beginning to understand how the technology works. And we still can't replicate it. The field of flying robots is advancing rapidly, but our efforts to replicate a dragonfly's flight capabilities in a model have so far resulted in something that is orders of magnitude bigger, slower, unwieldier, stupider, clumsier and harder to control than the real thing. Something, in short, that is nowhere near as good at being a dragonfly as a dragonfly is. Although, to be fair, they've had a 300-million-year head start.

Dragonflies might not be able to brew craft beer, throw a pot, write haiku, speak Polish, play the ukulele, make a 'To Do' list, find the end of the Sellotape, analyse sonata form, execute a perfect topspin lob, appreciate the elegance of a finely turned wooden bowl, devise holistic software strategies for implementing your long-range IT goals, or remove a stripped screw with a rubber band,* but they don't need to. They're dragonflies – masters of aerial agility – and that is, by any measure, quite something.

Some encounters stay with you.

They might not be spectacular or gasp-inducing. Sometimes it's simply the first one – your first peregrine, your first starling murmuration, the first time you notice a different butterfly skeetering about in the border. Sometimes it's the intimacy that grabs you – a moment of surprising eye contact, the feeling you get that, implausible as it might seem, you're looking into the soul of another creature. You see it in its entirety, and – more unnervingly – it sees you.

I have never had that precise feeling with a dragonfly – probably because of the nature of their eyes – but two dragonfly moments dwell in my memory, meetings that brought home just what we're dealing with when we come across these remarkable creatures. Moments that my mind keeps in a small recess and brings out whenever the word 'dragonfly' is mentioned.

Moment one. A summer's day. Warm, almost hot. Enough to make the English go pink and look flustered. There is sun, there is somnolence, there is water. Cool, fresh water. Ten per cent better for humans; essential for dragonflies.

* Imagine my amazed delight when I discovered that this basic DIY tip was actually a thing. Imagine, too, my frustrated rage when I realised that it mostly doesn't work.

Fffllrrreoooo.

So close I can hear it.

The sound – four wings flapping at about 20 beats per second – triggers the searching instinct, and after a few seconds I see a glint of red low down and find it.

Them.

They are – and there is no delicate way of putting this – at it. And in the world of Odonata that sometimes means flying around stuck to each other in a configuration that looks at best uncomfortable and at worst dangerous. Some damselflies form a sort of heart shape, connected to each other at both ends. This pair – common darter dragonflies – are forming a position I call 'The Tow Truck', the female's head attached to the male's tail. At the very least the female gives the impression she might be privately thinking, 'Owowow. Let. Go. Of. My. Nose.'

Quite how they manage to overcome the inevitable turbulence, vortices and wakes created by four pairs of wings flapping in close proximity is beyond me. Not content with that, they now execute a trick requiring such strength, control and agility it quite makes my head spin. Which is, I imagine, nothing to how it makes the female feel.

I get down on my knees, to their level. Some things are worth dusty trousers. The male, somehow keeping his own head still, flicks her like a whip, so that the end of her abdomen slashes down to the water, just barely touching the surface. *Flick slash flick slash flick slash.* All the while controlling their position in the air and moving gradually across the pond so that the eggs she is laying aren't all grouped together. Mesmerising behaviour, fixed in my mind's eye, replaying GIF-style on demand. All for the continuation of their species.

Moment two. A June afternoon at the Alhambra in Granada. We're there for the gardens, the architecture, the history. But some things catch your eye, no matter how splendid the surroundings. And the surroundings are splendid. Vaulted ceilings, mullioned windows, intricate carvings, tessellated tiles, decorations absolutely everywhere. Grandeur, intimacy, proportion, scale. Stunning views.

It continues in the gardens. Olive and cypress, oleander and box, wisteria and jasmine. Beautifully tended borders. Courtyards balancing light and shade, warmth and coolth. Everything in perfect equilibrium, satisfying all the senses. And everywhere, so pervasive and subtle that you barely appreciate its soothing effect, water. You can't heave a brick in the Alhambra without hitting a fountain.

There is a lot of the Alhambra, and it is well worth the trouble. But after a while the touristing takes its toll, and what you want is a nice sit-down, away from people, to let the afternoon drape itself over you. An aimless saunter leads to a small enclosure away from the main drag. I take a seat on one of the stone benches. It's that time of the afternoon when the world seems to pause. 'Sultry' is a word you might use to describe it, if so inclined. 'Arse-meltingly hot' are three more.

The pool in the middle is long and rectangular, low stone walls retaining the water, twin fountains at each end providing sight and sound restful to the senses. It would be enough just to look at the arc of the water and relish its splashy twinkling in the hot Granadan sun. But I soon become aware of another distraction.

A dragonfly.

I don't know what species it is; nor do I care. All I know is that it's big. It seems, to my inexpert eye, the size of a small bird.

It's doing lengths. Even against the dark backdrop of the hedge behind I can pick it out clearly, its wings occasionally glinting silver in the hot afternoon sun. There is something mechanical about it, not just in the whirring flight, wings moving too fast for any individual movement to be discernible, but in the regularity of its flight path. Sprint the length of the pond, pause, turn, sprint, pause, turn, like a swimmer in training. To think about it for even a minute is to wonder how it is done. Its control seems complete, each movement executed with purpose and certainty. You could imagine it as the product of the design department of a science fiction film. And you wouldn't want to be hunted by one.

It treats me to a ten-minute show, an aerobatic performance for which I would happily have shelled out twenty euros. But this is free of charge. And then, without warning, it swerves out of its regular path, darts over my head and away out of sight to more fertile hunting grounds.

I ponder it for a few minutes, asking myself questions about how and why and what and where and when. Perhaps prompted by my surroundings, I travel back in time, montage-style, imagining its distant ancestor, exhibiting the same behaviour over a stretch of water not far from here. I am surrounded by architecture that encourages us to think about a millennium of human history – but this is a timespan put into perspective by the presence of the dragonfly. Its flight, a wonder of physical and neurological coordination, evolved to make them consummate hunters and carries with it the memory of millions of years – a legacy we can't hope to match.

And it's not the only one.

3

THE BEETLE

The local park on a bright spring day. Grass, for lying on. A flowerbed nearby. There is freshness in the air, the tangible energy of growth. Insects are emerging.

One of them is a childhood familiar. It's crawling, apparently without a care in the world, across a rose leaf. Combined with the green backdrop of the leaf, its shiny red carapace and neat spots make a strong argument for the visual excellence and soothing properties of nature.

From long habit, without thinking, I say the first line of the verse.

'Ladybird, ladybird, fly away home.'

But I don't want to cause it distress, so the rest – the bit about burning houses and absent children* – remains unspoken.

We're well disposed towards ladybirds. Part of this lies, I think, in their appearance. They're an appropriate size for an insect – neither invisibly tiny nor terrifyingly large – and there is nothing remotely threatening about the elegant, shiny curve of their protective carapace. Any icky bits – antennae, proboscis, legs and so on – are either tucked away out of sight or small enough to be acceptable to even the most sensitive of dispositions. The design brief was clearly 'smart casual insects for lovers of pretty'.

This assessment might meet with howls of protest from the aphid community, which, along with thrips and mites and any number of other tidgers, is susceptible to the ravages of predatory ladybirds. But as any allotmenteer will tell you, we regard aphids as pests because they chomp things we like, so the ladybird's role in controlling their population is welcomed.

This one is doing no aphid-chomping, contenting itself merely with a quiet crawl across the broad expanses of the leaf. Then the shiny red shell opens up like a miniaturised DeLorean, and it pauses for an instant.

This transformation is always faintly surprising. Ah yes. Ladybirds fly. Of course they do. It's right there in the nursery rhyme – 'fly away home'. But the form we're used to – neat and compact, not a wing in sight – is unaerodynamic, apparently more suited to gentle bimbling than aerial manoeuvres.

* The origin of the hard-hitting – some would say grisly – rhyme is, like so many things, a source of much speculation. Some would have it referring to the persecution of Catholics, others farmers offering a friendly heads-up to the insects prior to stubble-burning.

In the instant before it takes off, I have time to appreciate its new form. The wings themselves are standard-issue insect wings, slender and wide, attached to the ladybird's rear portion. But that carapace has opened up at the front, and now resembles two oversized comedy ears. Shell-like and attractive, for sure, but not the stuff of flight. It looks like a first-round loser in a competition to design a flying machine, cobbled together more in hope of aerial prowess than expectation.

But what do I know? As if in defiance of my unspoken and disrespectful thought, the ladybird whirrs into action, rising smoothly from the leaf and up and away and out of my sight.

Draw a species from the tombola barrel of animal life on Earth. Three to one it's a beetle.

If insects are Planet Earth's diversity success story (and they are), then beetles are their flag-bearers. They number, at the last count, about 400,000 species. That's 40 per cent of insects, and 25 per cent of all animals. And there's a pile more in the in-tray, awaiting description. The actual number is probably at least a million, maybe more. And most of them – about 92 per cent – fly.

Put a random word in front of the word 'beetle', and you stand a fair chance of hitting on a species name: furniture, skin, lady, leaf, click, stag, dung, rove, clown, rain, betsy, hide, jewel, pill, riffle, water . . .

You get the idea.

A million species. Like a million years, it's more than the human brain can comfortably encompass. This abundance is supposed to have prompted the biologist J. B. S. Haldane to utter one of his more quoted lines – although most references to it (including this one) include the obligatory caveat 'probably apocryphal'. It arose from a question about the nature of a putative Creator of All Life on Earth. What, the questioner wondered, might we deduce about such a creator on examination of Their handiwork? Haldane's answer: that They must have had 'an inordinate fondness for beetles'.

As you'd expect with such large numbers, the diversity among the Coleoptera* is wide, ranging from the very tiny (the featherwing beetle – remember the name, as we'll be visiting it later on) to the gulp-inducingly large (the six species of Goliath beetle, whose larvae tip the scales at 100 grams – that's approximately the weight of a blackbird).† They live everywhere except the sea and the poles. Wherever there is vegetation, you are likely to find beetles. In appearance they range from the entrancing to the grotesque. For humans, they are significant: we eat them, worship them and wear them.

Exactly how long they've been around is – as everything in this field seems to be – a matter for vigorous debate. The consensus appears to be that the group under whose umbrella modern beetles shelter arose around the time of the griffinflies, a little less

* The word, as so often, comes from the Greek: *koleos* – 'sheath', *pteron* (we're beginning to get the hang of this) – 'wing'.

† Convention dictates that the weight of birds is expressed in terms of common coinage. The most commonly encountered manifestation of this is that the goldcrest, at 5 grams, weighs no more than a 20p piece. So the Goliath beetle larva is equivalent to a handy pocketful of change that could buy you a couple of coffees and a croissant.

than 300 million years ago, although the confidence interval for this figure is wide.[1]

While the griffinflies were bossing the air as hunters, those early beetles, the Protocoleoptera, went in a different direction – they are thought to have had a fondness for both boring into and eating wood. This served them well for a long time, but The Great Dying at the end of the Permian hit them pretty hard, mostly because it hit the forests pretty hard – wood-eating insects need wood, after all. But they bounced back. Diversification grew through the Triassic, and by the time of the appearance of the ground-beetle and water-beetle families around 225 million years ago, approximately 20 per cent of insect fossils are beetles, with all the four modern suborders represented.[2]

Cracking on into the Jurassic (approximately 201 to 145 million years ago), they became more adventurous in their diet. Now there were carnivorous and herbivorous beetles, including, from the late Jurassic, the Phytophaga – a huge clade,* comprising longhorns, leaf beetles and weevils, which is the second largest plant-feeding lineage of all insects, yielding only to the Lepidoptera.

The growth of herbivorous beetles coincided with a rapid and widespread diversification of flowering plants from the late Jurassic and into the Cretaceous (145 to 66 million years ago). Most of today's beetle families (almost 64 per cent) first appear in the latter. This conjunction is often closely implicated in their success, but it also helped that they were working from a solid base. An already lengthy evolutionary history with low extinction rates, combined with a natural adaptability to different environments and habitats,

* A group of animals or plants descended from a common ancestor.

helped establish a healthy springboard from which they could diversify once conditions were right.

This diversification was nipped in the bud first by the break-up of the southern landmass during the Cretaceous, and then by the pesky mass extinction – the Cretaceous–Paleogene (K–Pg) event – 66 million years ago that did for so much of life on Earth. The extent of this extinction – widely accepted as the result of an asteroid colliding with the planet – wasn't quite as large as the carnage caused by the Great Dying 185 or so million years earlier, but it looms large in our consciousness, not least because it caused the extinction of the dinosaurs. It also paved the way for the success of the mammals, so it's not surprising that we regard it as important. Surviving it was quite an achievement, but once the beetles had got it out of the way the true spread could begin, and they got on with the business that has resulted in the mind-boggling numbers cited above.

Four hundred thousand described species. It bears repeating.

So how did they do it?

There was the co-evolution with flowering plants. But they also underwent three physical modifications, each potentially contributing to their success.

The first change was to the muscles used to flap their wings. The direct muscle attachment system of the Palaeoptera, described in the previous chapter, is simple, intuitive and effective enough, but it has limitations. Most obvious is the upper limit on how fast the wings can beat – the contraction of the muscles is married to the speed of the nerve impulses (one contraction per impulse) and because the nervous system can't send another impulse until the activity from the last one has died down, the wings can only flap so fast. So insects using the direct flight mechanism beat their wings at no more than about 200 beats per second. To humans, that sounds impossibly fast, but in the realm of insect flight it's fairly pedestrian.

This limitation hasn't stopped the Palaeoptera from thriving, but beetles, along with all other insects (which amounts to about 75 per cent of them), use another, more involved method. Rather than having the muscles act directly on the wings, they developed a two-step process. Muscles deform the thorax, which in turn flaps the wings. Intuitively, this quasi-Heath Robinson arrangement seems more complicated than necessary. Surely it's easier just to pull the thing you want to flap? But the big advantage of this arrangement is that these muscles are antagonistic – that is to say, they set each other off, with a resulting increase in energetic efficiency.

The muscles are packed into the thorax in contrary directions: one set runs from front to back, the other from top to bottom. The first set contracts, squeezing the thorax, which pulls the wings down; and that squeezing triggers the contraction of the second set, which pulls the wings up; and that squeezing triggers the contraction of the first set . . .

And so on.

Above a certain speed (about 100 beats per second), these contractions aren't linked to the electrical impulses from the nervous system. They are, if you wish to give them the proper name, asynchronous. Possession of asynchronous muscles and an indirect flight mechanism means that an insect can flap its wings faster. Much much faster. For some tiny midges (called, appropriately enough, 'no-see-ums') this means in excess of 1,000 beats per second. Not that most beetles aspire to or achieve such extremes of velocity, but it's useful for some because of its efficiency, and the extra speed and mobility it potentially affords.

The basic wing system of the Palaeoptera had another obvious limitation: portability. The wings of mayflies and dragonflies stick out, immobile and unpack-away-able, when the insect is at rest. On the plus side, they're ready for instant use without recourse to tiresome unfurling. But they're more susceptible to wear and tear, more vulnerable to damage from predators, and less able to move easily in crowded environments.

There was another way. Beetles – among many others – developed a system whereby the wings flex back over the abdomen. Use them when needed, then stow them when at rest. It's such a simple idea, implemented by one tiny plate on the side of the thorax. This new space-saving technology presented an obvious evolutionary advantage, enabling insects to access enclosed spaces where they could shelter or roost. So once again, the mayflies and dragonflies and their relatives – with their adherence to outdated, non-flexing wings – had become the outliers. The Betamax of insects.

While these two changes were useful enough to ensure their adoption by the vast majority of insects, beetles made a further adaptation. It's their defining characteristic, distinguishing them from other insects – and it is very clever.

The typical insect has two pairs of wings.* Many use both pairs, to varying degrees and with varying levels of expertise, for

* If, at this point, you're sitting with your hand raised, bursting to say, 'But, Sir! But, Sir! What about flies, Sir?' . . . well, I sympathise, but I also point you to the next chapter, which deals with them.

flight. But beetles fly with just one pair. While the hindwings were reserved for flying, the forewings turned into toughened protective cases called elytra.*

The construction of the elytra is simple and ingenious. They meet in the middle, with a tongue-and-groove joint usually enabling a tight fit. Corresponding snugness around the outer edge is ensured by a slight folding under, like a fitted sheet. There are enormous variations in the shape, from almost completely flat to bulging domes, slimline to chunky. While the surface of, for example, a ladybird's elytra is almost entirely smooth, others might be bumpy, corrugated, ridged, ribbed or even covered with thin hairs. The variations seem almost endless – much like the variety of beetle species themselves.

Whatever their size, shape or appearance, the purpose of the elytra is simple: protect the hindwings. With the wings tucked snugly away under a rigid cover, beetles can crawl into the tiniest crevices, or burrow down into the earth, dung, leaf litter or rotten wood. A huge advantage in a dangerous world. And they offer general protection to the whole insect, too. Pick up a moth and you'll squash it; pick up a beetle and you'll merely annoy it.

The importance of elytra was demonstrated in a 2016 study in which large parts of the elytra of red flour beetles were removed and the results compared with undamaged insects.[3] The removal of the elytra universally resulted in lower survival rates and increased vulnerability to a number of factors, including predation, desiccation, wing damage and cold.

* From the ancient Greek *élutron*, meaning 'sheath'.

While elytra provide a major advantage, they also present all flying beetles with the same conundrum: what to do with them when they're flying. A few species – notably the attractive, metallic racing-green rose chafer – are equipped with a handy notch that allows the hindwings to flap freely while the elytra remain closed, and with this adaptation they're able to generate an impressive turn of speed. But most beetles open the elytra to release the wings before take-off, and there they stay, splayed wide. They do have a tendency to waggle or flutter in sympathy and synchronicity with the hard work of the flapping hindwings, but it's not entirely clear whether this helps or hinders aerodynamic performance. On the one hand, if they're big enough, the elytra can provide extra lift just by their very existence. But the configuration of elytra in front of flapping wings has the potential to create a mess of wakes and vortices that counter any potential advantage. If the arrangement is not noted for its streamlined efficiency, this matters little at the speeds attained by most beetles. And for many, flight plays a relatively small role in their lives. The most obvious advantage of flight is that it enables faster movement than other forms of locomotion, whether it's a short hop from one plant to another or wider dispersal to escape predation or to ease the strain on an overcrowded population. While predatory carnivores need a certain amount of agility and urgency to outwit their prey and thus tend to be among the stronger flyers, for wood-munchers, detritivores and herbivores, feeding is done while earthbound. Not for them the aerial hunting skills of a dragonfly – just get them over to that nice rotten tree trunk, thanks.

For all their advantages, the elytra also present new challenges for a beetle, not least the difficulty of packing the wings away. For an example of how this is managed, let us return – as if we needed any excuse – to our ladybird, just at the moment it decides to abandon me for pastures new.

The deployment of its wings is straightforward enough. The ladybird opens its elytra, and the wings snap into the stable and at least partially rigid position required for flight. This takes about a tenth of a second and is possible because the veins that give the wing the stiffness it needs to fly also have significant elasticity.

The reverse process is considerably more complex and time-consuming, and it's only recently that we've gained insight into exactly how they do it. These insights are the result of a 2017 study, led by Kazuya Saito of the University of Tokyo, which sought to unravel the invisible mysteries of ladybird wing-folding.[4] The mysteries were invisible for the inconvenient reason that ladybirds close their elytra before embarking on the folding process, so it takes place out of sight. Saito's team overcame this by creating a transparent elytron, which they grafted onto a test ladybird* so they could examine the process in detail. It gave the ladybird a bespoke, designer appearance, and the scientists the opportunity to work out what the hell was going on under there.

The task: to fold an open expanse of veined chitin into an enclosed space.

Here's how a ladybird does it.

The elytra close, leaving the hindwings sticking out behind like an untucked shirt. The ladybird then uses a combination of abdominal pushing and pulsing to nudge the wing into position.

* The ladybird seems to have been completely unaffected by the transplant procedure.

But the thing that boggles the mind – and I speak as one equipped with two hands and opposable thumbs who can still barely fold a paper plane – is the precision with which the ladybird performs the procedure. The intricacy of the crease patterns of the folded wings, each mountain and valley slotting into a specific position for the most efficient stowing in limited space, will be familiar to anyone who has ever tried their hand at advanced origami.

Helping it are tiny, complex structures that hold the wing in position while the ladybird prepares the abdomen for the next bit of nudging. It proceeds slowly, because it's not enough just to cram the wings willy-nilly into the cavity under the elytra, the way I would with a map, stuffing it into the glove box with a peremptory 'Ah, sod it, that'll do.' The wings are easily damaged and need to be used again and again, and the insect can't just go to Milletts and get a replacement. In particular, the veins, which not only circulate blood* around the insect's body but also hold the wing together, have to be treated with care and respect, and need to be a particular shape. If they folded sharply like paper the blood flow would be restricted and the veins would be damaged. So, their flexibility and structure enables them to maintain a cylindrical shape at the point of folding. This flexibility discounts the need for joints and other breakable moving parts.

The flexibility of beetle wings can also come in handy during flight. For those living in the wooded environments like those adopted by their distant ancestors, the risk of crashing into the surrounding obstacles is obviously increased. Anything living in such an environment is likely to develop a body plan to suit, narrow

* Yes, haemolymph – well remembered.

wingspans being the most obvious adaptation. Albatrosses don't live in forests. But many beetles have another trick that helps them negotiate the obstacle course. It's all in the wing-folding.

A 2020 study on rhinoceros beetles by Korean scientists Hoang Vu Phan and Hoon Cheol Park showed that if the beetle runs into something in flight, the wings partially collapse and then snap back into position.[5] This flexibility enables them to carry on flying in a crowded environment, despite the knocks and bumps from running into branches and other obstacles. Incidentally, the flexibility comes from our old friend resilin, of which rhinoceros beetles have an ample supply.

The researchers contrasted this behaviour with their own stiff-winged beetlebots flying through the same obstacles. Unsurprisingly, the slightest collision threw the beetlebot off, they couldn't recover and they crashed. So the researchers remodelled the wings, based on what they had learned from the rhinoceros beetle, and sure enough the beetlebot performance duly improved.

The rhinoceros beetle, as the name implies, is among the largest beetle species – measuring up to 15 cm in length – and usually chunky with it. While flight is aerodynamically within their grasp, some of those larger beetles are unable to take off from the ground, having to climb to a perch from which they can drop down and open their wings. When in flight, they hold their bodies vertically, looking for all the world as if they're suspended by an invisible hand, their fate at the mercy of the whim of the suspender. Their legs dangle below them, but while this might seem a hindrance to efficient flight, it turns out there's a reason for everything – they act as stabilisers, counterbalancing sideways movements and helping keep the ungainly creature on the straightish and narrowish.

A good example of this bumbling flight – visible on warm summer evenings if you're in the right part of the world – is performed by stag beetles. The males of these strikingly large insects develop equally dramatic mandibles – sometimes as long as the rest of the body. These antler-like appendages are put to good use in the mating season, the beetles engaging in aggressive battles to see off the competition and to impress females. Such is the potential for damage caused by the mandibles in these encounters, a stag beetle's elytra are correspondingly tough. They're multilayered and combine strength with flexibility – hard on top to protect against puncture from a thrusting mandible and flexible underneath to increase load-bearing capacity (although sadly not enough to protect them against the weight of careless human feet).

While the mandibles are good for fighting, they are undeniably unwieldy, to the extent that for most male stag beetles, flying is more efficient than walking. Around dusk they take to the air, on the lookout for a mate, their slightly meandering path giving the impression that they've just had a couple of large gins. There is something about them that says, 'We're not really supposed to be doing this, but if you absolutely insist.' Their progress is steady and unspectacular, often accompanied by a low spluttering droning sound that calls to mind the dogged flight of an old biplane. But they eschew the straight line for a reason. What they're after are pheromones – released by females and floating on the wind, as elusive as a phone signal on a remote island. The male zigzags here and there, hoping to latch onto whatever plume might waft into his sensors. His flight is about as far removed from the deftness of a dragonfly as can be. As with a dog playing the trombone, we marvel at it not because it is done well, but because it is done at all.

But at least he can do it. Some beetles, having developed wings, decided over the years that they didn't need them after all and became flightless. One such beetle gives an extreme example of the protective function of the elytra, the magnificently named diabolical ironclad beetle. This flightless, desert-dwelling beetle's shell is a complex, jigsaw-like structure based on interlocking elytra that cover the whole body. It is so astonishingly resistant to extreme forces that you can drive over it in a car and it will survive.

At the other extreme are the rove beetles, a prolific group numbering about 63,000 species, which have somehow managed to get by for a couple of hundred million years with shortened elytra, leaving their backsides exposed for all to see, and necessitating an especially complex and condensed folding pattern. But the shortened elytra make them even more expert at hunkering down into the tightest of spaces, whether under a log, into a rock crevice or any of the many places beetles shelter. They mostly fly effectively enough, too.

In many cases, the elytra have evolved a secondary role. A proportionally large, unimpeded external surface can be used for display. The appearance of beetles not only varies greatly in terms of shape and size, but in patterning and colour as well. Some are, to human eyes, strikingly attractive – jewel-like iridescence, rainbow-coloured patterns, spots, stripes, squiggly lines that look great on a beetle shell but you wouldn't want as a wallpaper pattern in your sitting room. One Central American species, *Chrysina resplendens*, has a shiny gold outer casing that wouldn't look out of place on a Lamborghini. Others look – to our eyes, but not, one hopes and assumes, to potential mates of their own species – grotesquely ugly. Beauty, eye, beholder and so on. Some beetles are aposematic – adopting bright colours as warnings to potential predators that they are poisonous. Others go in for Batesian mimicry, cleverly adopting similar gaudy colouring with a view to conning potential predators into thinking they are poisonous, even though they're not.

And on and on. Beetles all the way down, with more to come. In 1982, American entomologist Terry Erwin, using fogging machines and collecting sheets, recovered 1,200 beetle species from one species of tree (*Luehea seemannii*).[6] Of these, he estimated that 163 were exclusive to that tree. Extrapolating from this, he estimated that there were around 30 million arthropod species in the world, of which 12.2 million were beetles. While this estimate has since been revised downwards, Erwin's fogging technique highlighted the extraordinary abundance of life in the canopy, and just how much of it there is that we don't know about. Sadly, the likelihood is that innumerable life forms have gone extinct (probably as a result of our actions) without our even being aware of their existence.

Insects are the most abundant and diverse animals on the planet, and beetles head the field. This diversity is the result not of developing the ability to fly to its absolute limit, but of taking the tools for flight and adapting them to another purpose. Beetles might not all be brilliant, spectacular or eye-catching flyers, but they do it just well enough.

There's a lesson in there somewhere.

4

THE FLY

A short post-lunch walk. Making the most of the last day of the holidays. Butterflies flit. Buzzards soar. A yellowhammer darts to the top of a bush and treats me to a minute of sweet song. My eye is drawn to the pleasing fractal pattern of an umbellifer. It's the comparatively brutish hogweed rather than the dainty cow parsley, but pleasing to the eye nevertheless. Two small, narrow, orangey-red rectangles adorn the surface of the flower, locked in coleopteran passion. Red soldier beetle. They come out in midsummer and spend most of their brief adult lives having sex. Nice work if you can get it. They and their ancestors have been doing it for millions of years. I am somehow cheered by the thought.

As I move away, my foot nudges something. It's the corpse of a small bird – a meadow pipit, perhaps, although it is in the mid-stages of decomposition, when identification points are far from clear. A blizzard of flies – the big, thuggish kind – erupts from the cadaver. They swarm around my head for a couple of seconds before dispersing briefly and returning piecemeal to settle on the deceased. Or 'lunch', as it is now known.

I'm struck by one aspect of the episode. Flies, I realise, are incredible flyers.

This is hardly surprising – the clue's in the name – but there's something about being completely surrounded by something, however briefly, to make you notice it in a different way. Their nimbleness and speed are one thing – they also follow unpredictable flight patterns that bewilder the human senses, and they do it with absolute assurance. There's an aggression to it all that is far removed from the meandering flight of our early-evening stag beetle, and this impression, I realise, comes at least partly from the violent certainty of the flight. Taken purely as a feat of flying, what they've just done is astonishing.

Despite this feeling of awe, I would be lying if I pretended it didn't take the gloss off my walk just a tidge. Because the trouble with flies is this. We hate them.

It doesn't really matter which of the many species in the order Diptera (from the Greek for 'two wings') is under scrutiny, the human reaction is most often one of revulsion.* They're ugly, dirty, bearers of disease, destroyers of crops. At best they're just plain irritating. We're blind to any upside. When you're being bitten to hell and back by midges, you don't care how brilliant your assailants are at flying or how they do it – you just want them to go

* Strictly speaking, we should refer to them as 'true flies', as if all the other species with the word 'fly' in their name – butterfly, dragonfly, etc., etc. – are somehow pretending. Which, if you think about it, they are.

away. Then there's the buzzing. That low-level, incessant buzzing around the room, drilling into your head at a frequency perfectly calibrated to set your nerves jangling and – eventually – drive you mad.

We might make an exception for crane flies.* They're harmless – non-biting and non-stinging – and their gangly stumblings can be endearing. And once you've learned to distinguish hoverflies from bees and wasps, their almost preternatural ability to suspend themselves in the air is a source of fascination. But those exceptions aside, the history of our relationship with flies is beset with mistrust and unease.

At this point it's difficult to adopt a breezy 'but hey, come on now' tone and encourage people to look on the bright side. Yet we must always seek balance. Nature is complex and nuanced. In the plus column we have, for instance, pollination. Flies are brilliant at it, second only to the Hymenoptera (bees and wasps and their myriad cohorts) in terms of percentage of crops visited.[1] And pollination is widely recognised as Quite Important, so their role should not be ignored. Specifically, chocolate-lovers should give thanks to the genus *Forcipomyia* – a kind of biting midge that bears a great deal of responsibility for the pollination of the cacao tree.

Throw in the helpful habit of blow flies to tidy up – carrion, faeces, you name it – and then add the role some fly species play in keeping other so-called pests in check, and the ledger balances rather more evenly than many think. Then take a step back, think for a second about their size, assessing them from a purely engineering standpoint, and you might ask the same questions Pliny

* You might know them as 'daddy-long-legs'.

the Elder – that noted early observer and recorder of the natural world – asked about Mother Nature's ingenuity:

> Where is it that she has united so many senses as in the gnat? Not to speak of creatures that might be mentioned of still smaller size. Where, I say, has she found room to place in it the organs of sight? Where has she centred the sense of taste? Where has she inserted the power of smell? And where, too, has she implanted that sharp shrill voice of the creature, so utterly disproportioned to the smallness of its body?

Where indeed?

The word 'fly' contains multitudes. About 150,000 named species, to be imprecise, and probably a million or so in all, to be even more so.

There are tiny flies (*Euryplatea nanaknihali,* at 0.4 mm long, is reckoned to be the smallest) and big flies (*Gauromydas heros,* at 7 cm, the biggest). There are dainty ones, chunky ones, rangy ones, compact ones. Some seem to be nothing but legs; others are clearly fashioned solely from hairy menace.

But the species most people think of when they hear the name 'fly' is *Musca domestica,* the house fly. Our commensal, familiar to humans all over the world, its presence is so associated with our untidiness that you wonder how they managed before we came along. It is in fact a relatively recent arrival to the scene, thought to have evolved around 50 million years ago, but its ancestors date back to the Triassic – the first fossil is from about 250 million years ago, around the time of 'The Great Dying'. In the intervening time there have been three distinct radiations. First, the Nematocera ('long horns') about 220 million years ago – contemporary with the first records of ground and water beetles; then, hottish on their heels, the Brachycera ('short horns') about 200 million years ago – just ahead of the first true dragonflies; and finally the Schizophora ('split bearing'), 65 million years ago, shortly after the K–Pg extinction event had wiped out a load of competition.[2]

Like beetles, the Diptera have been very successful. Not least because – notwithstanding the existence of a few outlying wingless species, which, by being unable even to do the thing they're named after, besmirch the name 'fly' – flies fly. Flies fly really very well indeed. And the one thing that unites them – and the secret to the manoeuvrability of the most agile specimens in the order – is again what happened to their wings.

One pair of wings was quite enough for flying. But while beetles converted their spare pair to armour plating, flies took another path. 'What we want', they thought,* 'is manoeuvrability, coupled with insanely fast reaction times'. So they kept the forewings for flying, and shrank down the hindwings, converting them into something both very small and very useful.

What they turned into are tiny organs, shaped like miniature golf clubs,† called halteres. Like the pterostigmata of dragonflies,

* They didn't do it consciously, of course. It was all the result of evolution. But we're familiar enough with the concept by now to allow a bit of licence when describing it.

† Approximately a 7 iron, as far as I can tell.

halteres punch above their weight when it comes to their influence on flight. But while pterostigmata are more like a dragonfly's passive flying aid, halteres actively control the way a fly flies. Cut off its halteres, as early experimental scientists discovered, and a fly becomes flightless. You might as well cut off its wings.

Halteres act like a gyroscope. When a fly's wings flap, the halteres oscillate in time with them, and the knobs at the end swing up and down at extreme speed, like a demented pendulum. Apply rotation to them, and a phenomenon called the Coriolis effect comes into play. They bend slightly, sending signals to the steering muscles – in themselves a sophisticated set of springs that impart extraordinary control to the wings – to make any adjustments necessary. All this is happening about 200 times every second. The upshot is that flies can fly straight and true, even in the dark, correcting themselves faster than the blink of a human eye.[3]

The superpowers of the halteres aren't restricted to steering. They're a complex control centre, laden with sensory cells that give the fly information about the world around it and regulate the timing of the contraction of flight muscles with astonishing precision. Gyroscope and metronome, all in one.

This sophisticated and fast-acting balance and guidance system is just one part of a collection of top-of-the-range sensory equipment. Flies, like dragonflies, have compound eyes – their 3,000 or so facets might not be as impressive in number as the 30,000 sported by their distant cousins, but they're at least as fast. And they have hundreds of sensory cells on their wings and at other strategic places on their body, all sending important information to the brain and enabling aerobatic manoeuvres that are – with apologies to dragonflies – unsurpassed. Take, for just one example, their ability to land upside down on the ceiling. We see it, shrug and think, 'Yeah, fly on the ceiling – it's what they do', when we should, if we were properly curious and fascinated, be texting, 'I BEHELD A MIRACLE' to all our friends. Even when we see it slowed down, it seems weird and unnatural, a defiant 'fuck you' to everything we think we know about gravity. The fly flies directly up at the ceiling, daring it to back off. The front legs are the first to touch up (not down, obviously). The fly braces against them and executes a speedy backflip to bring the remaining four legs into position with an aplomb that garners unanimous sixes from the judges.

Maybe, as well as finding them repulsive, we don't marvel at these things because we don't see them. They're simply too fast. It's only when we slow them down to a speed we can absorb and process that we go, 'Ah yes. Impressive. Well done.'

It's not surprising that insects have quick reactions – their neural lengths are much shorter than those of larger animals, so the impulses take less time to reach the brain – but in the case of some fly species, this reaction time is optimised by sending information almost directly from sensors to wings, without processing by the brain. So that ultra-fast movement is purely a reaction to stimulus rather than a thought-out response to danger. The fastest startle-reflex times have been found in flies. In the house fly it's between 30 and 50 milliseconds, but the record-holder is a long-legged fly, clocking in at less than 5 milliseconds. If you've ever wondered how to catch a house fly, the secret isn't to try and move faster – that's a race the fly will always win – but to move extremely slowly. That way, you stand more of a chance of fooling it.

An experiment, in three stages.

Swoosh your arm through the air, as if playing a tennis forehand. Not only is swooshing your arm through the air a decent way of getting some mild aerobic exercise, it also, on this occasion, has scientific value.

The next stage is to go to your local swimming pool and swoosh your arm through the water.

The third stage, and the least easily executed, is to swoosh your arm through a vat of honey.

The most basic observation will yield the unsurprising result that stage one was easier than stage two, which was in turn easier (and much less sticky) than stage three.

It's all to do with viscosity. And this brings us to Reynolds numbers.

Osborne Reynolds was a nineteenth-century physicist who performed experiments in fluid dynamics – a subject guaranteed to send a chill down the spine of anyone unwise enough to attempt to understand its complexities in anything beyond the scantest of detail. But never fear; scant detail is all we need.

What he did was this. He added a jet of dye to water flowing through a large glass pipe and observed how the dye behaved as the water's flow rate varied. The jet held itself together at lower rates of flow, but as the velocity increased it dispersed. And there was a particular point at which the flow transitioned from smooth to turbulent. His experiments – changing the flow rate with a control valve at one end of the pipe – yielded the Reynolds number:

the ratio between the inertial forces and viscous forces acting on a body in a fluid.

If those words bring on a slight glazing of the eyes, all you need to remember is this: at low Reynolds numbers, flow is smooth;* at high Reynolds numbers it is turbulent.

Low, smooth; high, turbulent. It's not that snappy but repeat it twenty times while you're in the shower and it'll stick – it worked for me.

But what exactly does it mean?

An example. A Boeing 747 – very large and very fast – operates at high Reynolds numbers (typically in the hundreds of millions). The forces of inertia (basically its weight and its speed) far outstrip the forces of viscosity (the thickness of the air). The turbulence it causes in the air around it is correspondingly high. This is one reason why streamlined wing design, to reduce drag, becomes more important the faster an aircraft goes.

Working our way down the scale of size and speed, and switching to animal flight, a swift operates with Reynolds numbers in the tens of thousands; a wasp in the hundreds. The size of the thing travelling makes a difference, as does its speed. Lower either or both of those, and the Reynolds number goes down.

The transition from turbulent to smooth flow comes at a Reynolds number of about 2,000. Very roughly, as things get smaller, the air seems thicker to them. And as things get smaller, minimising the surface area of the wings becomes more important than streamlining if you want to reduce drag.

One of the useful side effects of the Reynolds number is

* Or 'laminar', if you prefer the word used by scientists.

scalability: if the Reynolds number is the same, things of different sizes will behave in the same way. For example, if you want to do experiments to determine the behaviour of fruit fly wings under certain conditions, you work out their Reynolds number, make a larger model with the same value, and experiment away, safe in the knowledge that the physics acting on the larger model will apply equally to the less manageably sized fruit fly.

When you reach the realm of the truly tiny, things start to get even more interesting and strange. Because insects smaller than about half a millimetre (give or take) sing to a different tune. At that scale, physics works differently.

Damn you, physics.

For an insect smaller than about a millimetre, the viscous forces take over to the extent that flying through air feels the way moving through honey would to us. And the appearance and operation of their wings changes as a result.

And now we briefly switch back, for the purposes of this demonstration, to the world of beetles. Specifically, the featherwing beetle, which is about 0.4 mm long. If the fruit fly – weighing less than a grain of sugar and with a wing area of just over two square millimetres – is small, then we can definitely apply the descriptor 'teeny' to the featherwing beetle.

For a long time it was thought that these tiny insects didn't fly, but were carried through the air on the wind, like thistledown. Again, technological advances have enabled closer study of how they move, and again it is high-speed camera work that has proved so enlightening. The beetle's eponymous feathery wings* reduce

* They're not the only 'feathery' ones. Thrips are an order of tiny insects, viewed by gardeners either as pests or pollinators (depending on the

its weight, and at this scale it doesn't matter that they're not solid aerofoils. In fact, it's better that they're not – their bristly structure has a much lower surface area, which reduces drag, and this allows them to buck the trend of 'smaller means slower', flying at the same speed as insects three times their size.[4]

As for the elytra, even at this scale they serve a secondary role, stabilising the beetle as it rows its way through the treacly air with an idiosyncratic flattened figure-of-eight wing motion. That wing motion, contrasting with the more straightforward back-and-forth action of most insects, is enhanced by a technique used by other very small insects: the 'clap and fling' method.

This technique, first described by Danish zoologist Torkel Weis-Fogh in 1973[5] and subsequently named after him, is how tiny insects overcome the difficulties of lift production at small sizes. Picture, if you can, a scuttle fly of the Phoridae family. This is the family containing the smallest known fly in the world mentioned above, *Euryplatea nanaknihali.* Typically about half a millimetre long, they resemble fruit flies, but are about a sixth of the size. I would, I think, struggle to see one if it were on my desk, and I certainly wouldn't be able to discern any details such as wings (their span is about the same as their body length) without a visual aid. As for the halteres, forget it – they're in microscope territory.

Strictly speaking, if adhering to the description first proposed by Weis-Fogh in his groundbreaking paper in 1973, the 'clap–fling' mechanism should be called 'clap–fling–flip', because those are its

species). Their scientific name, Thysanoptera, means 'fringed wings'. Like the featherwing beetles and a group of tiny wasps known as fairy-flies, their narrow wings are fringed with long hairs.

three stages. The wings meet at the top of the upstroke; they are then flung apart at some speed, before rapidly flipping 180 degrees to start the cycle again. Crucially, the clap is executed not the way we would clap our hands – flat surface to flat surface – but with a slight rotation, the wings rotating first around the leading edge and then around the trailing edge before flinging apart. The vortices created by this action generate the required lift without producing excessive drag, and the air expelled by the flung wings contributes additional thrust. The flexibility of the wings and the speed of execution also play their part in making this an effective technique, although our tiny fly pays a price for this. It turns out that clapping your wings together at high speed – upwards of 200 beats per second, if we needed reminding – can damage them, compromising speed and especially manoeuvrability over time, although not to such an extent that the technique is abandoned altogether. Wing damage is an occupational hazard for all flying animals, but wings are made to be more effective than necessary, so the buffer zone between being able to fly and being rendered flightless is larger than you might assume from looking at an insect's apparently fragile wings.

Whether using the LEV or the clap–fling–flip method of power production, all these insects need some way of controlling the direction of flight. For this a fly has sets of tiny steering muscles at the base of the wings that can control the angle of the wingbeats. By changing the moment of their activation in the stroke cycle even a small amount, they can generate subtle changes in the movement of the wing. This enables them to combine the six degrees of freedom that act on any flying body, familiar to anyone who has ever tried to fly any kind of remote control drone: the translational degrees – lift, thrust and sideslip – and the rotational ones – yaw, roll and pitch.* Of all flying animals, none has more control over their position in the air than a fly, and none can manipulate it with such ease and speed. And all with mechanical and neurological hardware that outstrips by a large margin the most sophisticated miniaturisation yet devised by humans.

That control can come in handy when tracking down food. Of all the very small flies, it is perhaps the fruit fly† (*Drosophila melanogaster*) that is most familiar to us. One minute the fruit is sitting in the bowl, not quite ready to eat; the next there's a tiny blemish on the skin; and before you know it the bowl is swarming with squadrons of the little buggers. And the same goes for that glass of fruity red you've just poured.

Our senses are dull. We do not have advanced smelling equipment. Even those well endowed in the olfactory department lag far behind the fruit fly. But then we don't have sensitive antennae attached to our heads, the better to detect a waft of chemicals at minuscule concentrations. The antennae are also handy for detecting the strength and direction of the wind in flight, but for our purposes, for now, they are merely scent detectors.

Here's how it happens. The fruit starts, ever so slightly, to ferment. The fruit fly, sitting around minding its own business, gets a whiff of fruity goodness – and off it sets. But the scent does not form a solid wall, or a predictable trail that the fly can follow all the

* Yaw revolves around the vertical axis; roll revolves around the nose-to-tail axis; pitch revolves around the wing-to-wing axis.

† I must apologise to any dipterists who happen, poor things, to be reading this. Because the creature known popularly as 'fruit fly' is, I gather, more properly known as 'vinegar fly'.

way to its source. The scent is fickle, floating around the atmosphere in wispy plumes, subject to the whim of any prevailing breezes. No sooner has the fruit fly found a plume than it might lose it. But if that happens, the fly changes its behaviour, zigzagging around until it can latch onto it again in a strategy not dissimilar to the romantically inclined stag beetle of the previous chapter, and the one I use when looking for that pair of scissors – I know I saw them just ten minutes ago; where the hell *are* they? – but much more effective.

Even better, when it finds itself in the presence of an attractive aroma, it scans the surrounding area, looking for shapes that contrast with the general milieu.[6] This is surprisingly canny behaviour for something with so few neurons (a mere 100,000 compared with our 100 billion or so, although when you factor in the comparative sizes of the two brains the fruit fly comes out ahead of the game). And it's helped by the same apparatus – wings, halteres, asynchronous muscles and all the rest of it – that allow the house fly to zip around so nimbly.

But there's an element of this that still befuddles the brain – scale. Take muscles. When we read that the fruit fly has a set of about seventeen muscles at the base of the wing acting together with the main muscles to control the insect's direction and manoeuvrability, we might take a moment to check the size of the fruit fly – 3 to 4 mm – and imagine how big those muscles are going to be. Then, like Pliny the Elder, we have Questions. Where does everything go? How does it do it? What sorcery is this? As with the enormities of geological time, our human-centric brains just need a while to get used to the idea of scale of small things. The notion of minuscule organs doing sophisticated things tests our imaginative faculties to the limit.

While each of the fruit fly's individual components are indeed worthy of attention and at least a respectful inclination of the head, it's the way they work together that makes them so effective. Taken separately, the antennae, eyes, wings, muscles – and especially the organs-formerly-known-as-wings, aka halteres – are remarkable adaptations. Put it all together and harness it to the control centre of the brain, and you have something rather special.

There's one in front of me right now, wafting around between me and my computer screen. Its flight path doesn't seem to have been planned in advance – it's just bimbling around, apparently on the off chance that something good might turn up. Somewhat against my better instincts and knowing that the four rock-hard pears in the fruit bowl could turn to squish at any second, I wish it every success in its endeavour.

Knowledge can be liberating.

For fifty years a buzzing fly in the same room as me would trigger irritation or disgust. Whether it was flying around pointlessly at the top of the room or hanging around near my food with a view to vomiting on it, my first reaction would be to look for a magazine so I could squash the bastard.

Nowadays, having learned something about their history, lives and habits, I pause to admire its aerobatic brilliance, thinking about its impossibly sophisticated guidance mechanisms, its tiny muscles, the processes that are taking place at a speed and scale almost beyond human imagination, and I take a few seconds to marvel at the processes that could have produced such an immaculate flying machine, and to ponder the extraordinary variety of life on this precious planet.

And then I look for a magazine so I can squash the bastard.

Not that I succeed. They are, as previously advertised, fast and manoeuvrable. Simpler, really, just to open a window and try to usher it out. It takes the hint and departs, halteres going at two hundred to the dozen, a flying advertisement for the benefits of taking an already successful body plan and tinkering with it. Two wings good? If backed up by important adaptations that bestow an advantage, certainly. Flies and beetles took completely different courses when repurposing their second pair of wings. And their contrasting approaches proved equally effective.

But while they were spreading across the world and occupying their own niches thanks to their original and daring decision to dispense with half the reason they got in the air in the first place, many other insects were emerging and succeeding with the more conventional four-wing configuration. Among them, a group containing both our favourite and least favourite stripy buzzers.

5

THE BEE

There's peace to be found in my little corner of South London. Away from the traffic, away from power tools, away from the guy over the road who plays industrial hip hop every Friday night. The local cemetery is a haven. I feel the cares of the world slipping from my shoulders as I walk through the gates. The noisiest things here are the parakeets.

There are carrion crows, extracting maximum nutrient value from an overflowing bin. There's a fox, trotting confidently across the path in front of me, pausing briefly to give me a curious glance and then moving on when it becomes clear that I'm not going to feed him.

And there are bees. A lot of bees.

I hear them first. A low, ambient buzz, somewhere behind me. And then, turning to look, I see them – a loose cloud heaving into view from behind the crematorium and inching doggedly across my sightline towards destination unknown, the buzz intensifying as they get closer.

My mind immediately turns to a bear of little brain, clinging to a balloon, mistakenly hoping to replenish his stocks of honey.

Unlike Winnie-the-Pooh's, these are the right sort of bees for honey, but just at the moment they have other things on their collective mind. I watch them pass overhead, trying and failing to track the progress of individual bees within the swarm. The buzz dies away, and soon it's just me, the crows and a vociferous great tit.

We might hate flies, but our feelings towards bees are generally more benevolent. We prefer it if they don't sting us, of course, and we like shouting at them when they find their way into the kitchen and then, rather than leaving the way they came in, opt instead to bash their heads against a closed window for half an hour – *the idiots*. But we like their honey and their wax, and their other by-products – among them royal jelly and propolis – have long been alleged to have magical health benefits. Plus, they are famous pollinators. So much so that plenty of people seem to think they're the only ones that do it. Our old pal Pliny the Elder thought bees 'alone, of all the insects, have been created for the benefit of man'.

Bees, in short, are fine with us.

There is also a tendency (and one I was guilty of for far too long) to lump the great variety of bee species (20,000 or so) under a maximum of maybe three general umbrellas: honey, bumble, other. This, as you swiftly learn once you start to delve into the subject, is grossly simplistic. On discovering, a few years ago, that there are 270 species of bee in Britain – within them varieties of nomad bee, carpenter bee, plasterer bee, mason bee, shaggy bee, pantaloon bee (definitely my favourite name) and plenty of others – I had to have a bit of a sit-down.

Bees belong to the order Hymenoptera, where they rub shoulders with sawflies, ants and wasps. Following a pattern that is rapidly becoming familiar, there is a bewildering profusion of species within the order – about 150,000 of them have been described, and some estimates would have them outnumbering even the Coleoptera. And they've been going a long time – the oldest known hymenopteran fossils are from the Triassic, 224 million years ago, making them more or less contemporaneous with the Diptera.

Much as we like to regard bees as benevolent pacifists and wasps as malevolent bastards, the former actually evolved from the latter. But the evolution into today's groups was a drawn-out affair, with the first wasps appearing in the Jurassic (201 to 145 million years ago), and ants and bees in the Cretaceous (145 to 66 million years ago). The sawflies are not only the earliest but are outliers, the other three groups being united by their extremely slender waists, which evolved as a feature relatively early on, in the middle Triassic. With apologies to sawfly fans, it is ants and wasps that catch the eye as a point of comparison with different aspects of bee flight and behaviour.

By the time of the emergence of bees, a lot of fossils were preserved in amber – fossilised tree resin, which we know as an attractive, often honey-coloured, semi-precious gemstone. An insect preserved in amber will often show a remarkable level of clarity,* making it much easier to examine attributes such as an insect's wings, and in hymenopterans these are a key distinguishing feature. Not for them the intricate vein network of the dragonflies, the hardened shells of the beetles, the whirring gyroscopes of the flies. They retain their four wings – sparse-veined affairs, fore significantly larger than hind. What sets them apart is a row of little hooks – hamuli – on the front edge of the hindwing. In flight, these attach themselves to the folded rear edge of the forewing, effectively turning four wings into two larger ones, and making them more effective aerofoils.† A neat trick if you know how.

Apart from the honey and the wax and the pollen, there was one other thing I knew about bees when I was a child. One of those 'facts' that went around, along with 'When you sneeze you're closer to death than at any other time in your life' and 'In your lifetime you'll eat ten tons of dirt.'

Bees are too heavy to fly.

There were no more details attached to the factoid. Just 'Bees are too heavy to fly.'

My immediate response was, I suspect, the same as most people's.

'Umm . . . so . . . how . . .? I mean, because I just saw . . .'

But the fact-giver was immune to such cheap barracking. A knowing look would come into their eye, and they would hastily change the subject.

* Sadly, the extraction of its DNA, the central plot point of Michael Crichton's *Jurassic Park*, remains science fiction.

† This coupling might be the origin for the name 'Hymenoptera' – a reference to Hymen, the Greek god of marriage. Alternatively – and less poetically – it's thought it might derive from the Greek for 'membrane wing'.

But the question remained with me. Are they 'too heavy to fly'? And, if so, how come you see them literally flying, like, all the time?

It didn't help that the myth was repeated at the beginning of the 2007 Dreamworks film *Bee Movie*, spreading the misinformation to several more generations of impressionable fact fans. The lie was already halfway round the world, while the truth not only hadn't started putting its trousers on, it was lying sprawled on top of the duvet in its underpants with an empty bottle of vodka clutched in its right hand.

Where the idea came from is a matter of anecdote. There are several versions of the story in which a drunken engineer jots down some calculations on a napkin/envelope/scrap of paper and comes to the astounding realisation that bees/wasps/insects shouldn't be able to lift their bodies off the ground. But for documentary evidence of the story's origins we might look to Antoine Magnan's 1934 book *Le Vol des insectes*, in the introduction of which we read, 'Inspired by what is done in aviation, I applied the laws of air resistance to insects, and, with my assistant André Sainte-Laguë, reached the conclusion that their flight is impossible.'

Quite the claim. And, it turns out, quite wrong. Magnan's hypothesis treated insects as if they were dictated by the same aerodynamic principles as aeroplanes. On the one hand this seems entirely reasonable – both are trying to make their way through the air, and both are subject to the laws of physics. But there is a huge difference between a heavy, fixed-wing metal tube and a box of chitin and muscle with mobile and flexible wings. That much is obvious, but more to the point there is a huge difference in how they each behave when moving through the air and how the air behaves around them.

Magnan's calculations told him that the amount of lift an insect could generate with its small wings wasn't nearly enough to haul its body off the ground. He was assuming they flapped them up and down, not back and forth, and didn't factor in what we now know about leading-edge vortices and so forth, which account for the magnification of lift produced from wings that on the face of it don't seem up to the task.

At what point the 'insects' of Magnan's book turned into the bees of childhood folklore isn't clear, but when we examine the way bees fly, we uncover something even stranger than the apparent defiance of the laws of aerodynamics.

Let's take, as an example, the western honey bee (*Apis mellifera*), flying advertisement for the benefits of the domestication* of insects, and familiar to millions from its catchy hits 'Waggle Dance', 'Big Swarm' and 'Has Anyone Seen My Queen?' My October bee, making the most of the flowering ivy, was a western honey bee, and the odds are that there is a colony somewhere near you. They're successful animals, mostly because of our partiality to their sweet, sweet by-product. Even if you haven't knowingly seen one recently, you're probably aware of their general appearance, and if asked, 'How do they fly?', you might, like me, give a little shrug before saying, with quizzical wrinkled brow, 'By flapping their wings?'

While their flight came to be much more thoroughly understood than in the 'they're too heavy to fly' era, research led by Michael Dickinson at Caltech and the University of Nevada in the early 2000s shed new light on its complexities.[1]

* Or at least partial domestication – opinions vary as to the extent to which honey bees can be called 'domesticated'.

General principles show that smaller insects compensate for their lower aerodynamic efficiency by flapping their wings more frequently than larger ones. This is fairly intuitive. Nobody would expect a cabbage white butterfly to flap as fast as a midge. But the 'tiny = fast, big = slow' rule of thumb doesn't apply to honey bees. Bucking the expectation for insects their size, they flap their wings at about 230 beats per second. This is a similar frequency to the fruit fly, made all the more astonishing by one crucial difference: they're about eighty times bigger. They manage such speeds not just with the sophisticated neurological network already referred to in flies, but by adopting a shallow wing stroke of about 90 degrees. When burdened with the often significant extra weight of pollen or nectar, they don't increase the speed of their flapping – they're already at the limit – but make their wing strokes deeper. This is right up there with the famous weight-carrying abilities of termites and ants – but in the air, backwards and in heels.

Indispensable though they are for flying, wings often have a secondary purpose.[2] Observe a hive closely, and you might see bees clinging to the entrance. This is a sign that the temperature or carbon dioxide concentration inside the nest has gone up, with potentially detrimental effects for the colony. When levels reach a certain threshold, the bees gather and fan their wings to move air through the nest and bring the temperature down – it stabilises at just under 34°C. Significantly, they use a much lower flapping frequency for this behaviour (approximately 175 flaps per second) than they do when flying – understandable when you consider that they're doing it not to lift their own weight but to move air, which requires considerably less effort. And they must of course be careful to remain still. The downside of flapping your wings near hard surfaces is the likelihood of wing damage, and even though their wingbeats remain shallow when ventilating, they're still vulnerable to the odd bang and scrape. This is an occupational hazard that the bees are able to accommodate, the benefit for the colony outweighing the inconvenience for the individual.

The complexities of colony behaviour in honey bees are many and varied, but for people of a certain age the most familiar of them will be known from over-exposure to 1970s schlock horror movies. The possibility of swarms of killer bees descending on rural Oxfordshire to deliver agonising death to me and my family was enough to scare me rigid. I'm sure I'm not alone. Sometimes the film-makers made an effort to explain this psychopathic behaviour. The bees were subject to some kind of mutation, or they were under the control of an evil human. But it's fair to say that rational, hole-free plotting wasn't necessarily a high priority for some of the films at the lower end of the genre, so 'just because' was also considered an adequate motivation. It seems curmudgeonly to carp about accuracy in such movies – a typical strapline was 'They're coming this way . . . not to make honey, but to kill' – but the idea that this is normal apian behaviour seeps into the consciousness, and when you have a population inclined to view all black-and-yellow buzzy things with suspicion, these impressions are difficult to dispel. For many people the news that a swarm of bees has been seen in their neighbourhood is enough to prompt a frisson of nervousness.

This all ties in to human fear. On the face of it, it might seem illogical that something as large as a human can fear something as small as an insect. But there are sound evolutionary reasons to be wary of an animal capable of delivering a single shot that might hurt or even kill you. And when they can fly, that fear is compounded – especially if, as so often, they fly unpredictably. A single insect might be a source of mild trauma; multiply it by a hundred, a thousand, ten thousand or more, and then make it mobile, and fear can become terror and panic. But at the heart of this is a misunderstanding of what swarms are, how they behave, and what their purpose is.

Take yourself to a warm day in late spring or early summer. The honey bee colony is thriving, resources are abundant and conditions are perfect for expansion. When the way has been paved for a new queen to take over the existing hive, about half of the colony take the old queen on a quest for a new venue – the longest and most adventurous single flight of their lives. The first stage is merely to leave the existing hive and park in a cluster somewhere nearby while scouts – about a hundred of them – form an advance party. Their mission, should they wish to accept it, is to find a suitable location for the new nest and then lead the swarm to it. Once they've left the natal nest, time is a major factor. They subsist only on the food in their stomachs and having left the old nest they can't return for fear of disrupting the building of the new colony. They have just a few days to find a new home.

The scouts head off, find a range of possibilities, then return to the cluster – sitting patiently on a tree and making sure the queen is protected – to perform the famous 'waggle dance', which communicates to their compadres the merits or otherwise of the selected venues. A decision is made, and the cluster heads off, earning itself in the process the new title 'swarm'. They must go directly there – potentially a distance of several kilometres, over hill and dale – as quickly as possible without anybody getting lost.

Nobody gets lost. The swarm, more than 95 per cent of which hasn't been to the new home, finds its way. What's more, as it gets close, it slows down so it doesn't overshoot, instead pulling up to the front door as smoothly as a stretch limo outside the Savoy.

The scouts are pivotal to the success of the mission. They're the ones who know where the new home is and must use their knowledge to guide their cohorts in the right direction. They do this by streaking through the top of the swarm with speed and purpose, to show the rest of them the righteous path. They go at the top to make themselves more visible to the group, silhouetted against the sky. Once they reach the front of the swarm, they slow down, drop to the bottom, allow themselves to fall to the back, then go again. With this rolling relay, the swarm is kept on track at all times. They also use their wings for pheromone dispersal to guide the rest of the swarm to the entrance of the new nest. They raise their body to expose the Nasonov gland, which resides at the tip of the abdomen. Out comes the pheromone, flap go the wings, waft waft, job done. The scent is dispersed and provides a trail back to the nest's entrance for other bees to follow.

It should be noted that at no stage does this process include any form of human attack. If someone were to jiggle a sitting cluster with a cattle prod, it might be a different story, but the bees are focused on finding their new home, so have little spare energy for extracurricular activities.

If killer bees were a source of childhood trauma, this was almost matched by the prospect of an encounter with their hymenopteran cousins: ants. Also narrow-waisted, also descended from a lineage of stinging wasps, ants are more closely related to bees than many people imagine, possibly because we encounter their flying form relatively rarely. The ants of my imagination were always poisonous and came in marauding armies that marched relentlessly over all terrains, their sheer numbers overwhelming paltry human efforts to stem the tide. All in their dogged quest to bite me – specifically me – to oblivion.

It's a good job I didn't know they fly.

The concept of 'flying ant day' is familiar to many people, especially tennis fans. It disrupts proceedings at Wimbledon most years, much to the amusement of the (frankly quite easily amused) Centre Court crowd. But 'flying ant day' is a misnomer, because it's not just a single day but a season, usually lasting from June to August.

They like it warm. They like it still. There's some evidence they like it when it's just rained. Whatever the trigger, for most people 'flying ant day' is a mild and temporary nuisance. For others it's a neat excuse to get on their local Facebook page and ask, 'What is it with these effin' ants??!!' And for a few enlightened souls it's an opportunity to observe ant behaviour at close hand. Though, admittedly, if you're caught in the middle of a swarm,* the phenomenon can be overwhelming.

You might get an early sign that the ants have taken to the air by observing the behaviour of gulls. They like an ant, do gulls, and

* The pedant will point out that the word 'swarm', correctly used, applies only to honey bees. But it's now so universally used for other flying insects that I fear the pedant's battle is lost.

when the air is filled with them, the birds perform mid-air gyrations worthy of an experimental dance troupe in their efforts to gorge on the sudden appearance of an all-you-can-eat buffet.

As with bee swarms, the motivation for this burst of aerial activity is dispersal – a straightforward way to diversify the gene pool. When the time is right, the queen switches – instead of laying female worker eggs, she lays male drones and virgin queens. Once hatched, these ants grow wings and take to the air in huge numbers, the idea being not only the usual 'safety in numbers' strategy but also to give them a better chance of meeting a mate from another colony. The queens mate with as many males as they can in their brief time in the air. For the males, this is their only purpose in life, and once they've flown and mated, they die. For her part, and showing admirable pragmatism, the queen will chew off her own wings and begin the search for a suitable site to set up a new colony.

Ant and bee swarms, despite our paranoid fears, are mainly benevolent. We might be relieved that they haven't taken inspiration from locusts. Unrelated to the hymenopterans except in the loosest 'also an insect' sense, they are nevertheless worth brief examination because of their simply extraordinary swarming abilities. Usually solitary, they undergo a genetic shift when conditions align, and become highly gregarious animals with an urge to breed en masse and swarm. And when they swarm, they properly swarm. Stories of millions of locusts blocking out the sun are as old as we are. They were, after all, one of the ten plagues visited on Egypt by God:

> God told Moses to stretch out his hand over the land of Egypt to bring a plague of locusts. The locusts covered the

> face of the land and swallowed up every crop and all the fruits of the trees. Afterwards there was nothing green in the trees, and all the crops in the fields had been destroyed.

This isn't, to judge by multiple accounts over many centuries, hyperbole. And these nomadic swarms still happen from time to time, presenting as much of a threat to the livelihood of farmers as they ever did, as evidenced by a series of massive swarms in East Africa in 2020. After a period of drought was followed by abundant rainfall, vegetation grew rapidly in the wet, sandy soil that locusts prefer, and the conditions were perfect for a breeding surge. One swarm in Kenya was reported to cover an area of 2,400 square kilometres. They can travel up to 160 km a day with the help of prevailing winds. Swarm, move, eat. And they do eat – approximately their own body weight on a daily basis. Trillions of insects descending on farmland can take out entire crops with devastating speed.

A swarm can be a rather more low-key gathering. Spiralling columns of nameless tiddlers in the garden of an early summer evening constitute a swarm of sorts. And while some insects swarm for specific reasons – breeding and feeding, as so often, come top of the list – sometimes the formation of a swarm is simply the result of a lot of insects of the same species being in the same place at the same time. There's no particular purpose to these swarms – they just are.

It'll come as a relief, after all this talk of nightmarish gatherings, to learn that wasps – the most reviled members of the Hymenoptera – are more unwilling swarmers. Which isn't to say they don't do it at all – just try attacking an active nest and see how you get on. But for social wasps, swarming is a defence against threat rather than a population dispersal strategy. The relative rarity of swarming in wasps, however, might be the best thing many people have to say about them, such is their lowly position in our estimation.

Plutarch suggested that wasps were degenerate bees, and this assessment prevails to this day. The general perception of wasps as useless and vindictive is deeply ingrained, largely based on the very few species we encounter on a regular basis – the yellowjackets and similar types that hang around our picnics and, so we think, sting us without the slightest provocation. Inconveniently for this anti-wasp narrative, the truth is (as so often) more nuanced than that, not least because there are so many species of wasp, the large majority living their lives away from the human gaze. It's true that some of them lead what human sensibilities consider unsavoury existences. The many species of parasitoid wasps* have habits that are not for the faint of heart, involving – as the word 'parasitoid' implies – mostly the infiltration of a host organism, leading to its eventual death. Placed next to this kind of behaviour, the perceived annoyingness of yellowjackets seems almost mild and eminently forgivable.

Almost.

Bees and wasps developed their sting not, as we would like to believe, as a form of unprovoked attack, but as a defence against predators, and, like it or not, that is exactly how they perceive us. We are, after all, the very large and often flailing monster stopping

* A lot of them are really very tiny, including the smallest recorded insect in the world, the wingless and blind *Dicopomorpha echmepterygis*, which, with a body length of just 0.127 mm, is smaller than a paramecium, the single-celled organism so beloved of biology teachers in the 1970s.

them from settling down on the picnic table and having a good old slurp of that blob of ketchup. Of course they're going to feel threatened by us. The sting is a sort of last resort – they've tried flying around in fast and unpredictable patterns, and yet still we flail. If they're lucky enough to get close to us without being splatted, the sting might just be the thing that sees us off. And when a stinging event takes place, they summon help.

Bee or wasp stings are, for most people, no more than painful inconveniences.* There is, however, one crucial difference. When you're stung by a female honey bee (the male drones don't sting), the sting, equipped with strong barbs, remains embedded, while the bee is eviscerated. A wasp retains its sting and is able to attack again. But in both cases a pheromone is released to summon the hordes and attack the enemy (that's you), and this is when a gathering of bees (not strictly speaking called a swarm in these circumstances) does pose a threat, because if you happen to be near a hive, this could be bad news. Honey bees aren't the fastest in the air, but they can reach speeds of up to 25 or 30 km/h – plenty fast enough to outpace a human. In this case their flight path will be admirably direct. Bees summoned by an attack pheromone, unlike a swarming colony relocating to its new home, complete the journey without passing 'Go' or collecting £200. We might find a moment to admire their efficiency, had we the leisure to observe it.

Meanwhile, we console ourselves with the fact that this scenario – given that most people are stung by a bee or wasp no

* If you're allergic to stings, it is of course more serious than that. It's estimated that less than 0.5 per cent of the population suffer anaphylaxis brought on by Hymenoptera stings.

more than once every ten or fifteen years – is more worrying in theory than in practice.

Unless you're the star of a 1970s schlock horror movie.

Enough of swarming. Enough of stings. Enough of tiny, malignant insects eating other creatures from the inside. Perhaps it's time to deal with something more benevolent, calming and comforting. And what could be more benevolent, calming and comforting than a butterfly?

6

THE BUTTERFLY

I'm never quite sure what makes me look. Maybe an unnamed instinct tells me that something's changed in the continuum; maybe my brain subconsciously registers a tiny movement; maybe the butterfly calls my name at an ultra-high frequency.

Whatever the reason, I look down and to my left, and there it is, sitting on the grass. A scrap of orange and black, the patchwork pattern on its splayed wings enhanced by tiny blobs of white near the edges.

It's a painted lady butterfly. One of the markers of British summer.

It is, I'm guessing, knackered. Long journeys can do that to you. And this butterfly is completing one of the longest relays, started a few months earlier by its great-great-great-grandparents.

Whenever I see a painted lady butterfly I think of a Peanuts strip from 1960. Linus and Lucy come across a butterfly – a big yellow one. Lucy, always keen to show off her knowledge and educate her brother in the ways of the world, tells Linus that it's flown up from Brazil. Linus crouches down, examines it, and declares that it's not a butterfly but a potato chip.

Lucy improvises masterfully.

'Well I'll be! I wonder how a potato chip got all the way up here from Brazil?'

How indeed?

The story of the painted lady's migration seems no more likely than the potato chip's would be. It starts, perhaps, in Africa – Ghana, say. The recently hatched butterfly sets off across the Sahara, fluttering gamely northwards towards southern Europe. The trigger for the start of this journey isn't exactly known. Maybe it's the temperature, or the amount of daylight. It's thought that the developing larva might respond to signals sent by its dying host plants – get out while you can, because we're not going to be around for you to eat much longer. Whatever the reason, the young adult strips the plant bare and moves on to pastures that will shortly be more abundant. This one, our representative sample, is strong, flying fairly low – low enough for it to be seen by an observer on the ground. On it goes, covering about 150 km a day, until it reaches Spain – the small matter of 4,000 km or so – where it stops, finds a mate, breeds and dies. Finding a mate is not hard, because they're all at it, a large population moving more or less in synchronicity. The turnover is quick – a whole generation, from laying of egg to death of adult, probably lasts between five and six weeks – and the next-generation butterfly, once hatched, continues the journey northwards. On, on, on. Breed and move. Breed and move. Some stop in France. Others crack on northwards, to

England, maybe even Scotland. Rarely for butterflies, they have been recorded in Iceland.

And then what? The return of the migrating population isn't necessary for the continuation of the species – not all butterflies in a population migrate – but it is necessary for the continuation of the migratory habit in the species. For a long time it wasn't clearly understood. It was thought that the migrators reached northern Europe and died out, unable to withstand the harsh winter, while the source population stayed put. But science makes new discoveries all the time. So now we know they make the journey south again in autumn.[1]

But those southward-travelling butterflies are more likely to adopt a different flying strategy. They go up. High up. Up to, perhaps, 500 metres, where, with the wind at their backs, they can travel south at 50 km/h. Where their predecessors made the northward journey over four or five generations, the butterfly on the return leg – which might be six generations removed from the one that took off eight months earlier – will complete it by itself. The height and speed of their journey, without stopovers on the ground where they might be found by researchers, meant that these butterflies went undetected for many years. It was only thanks to an international monitoring project, using specialist radar, that this part of their journey was discovered.

Reading about this multigenerational odyssey, I have questions. For example: why? If travelling is so exhausting and dangerous – and it is – why not just stay where you are? Plenty of butterfly species do adopt this strategy. They stay in the same place, feed up, and counter the inevitable difficulty of winter by going into diapause – a state of suspended animation – before emerging to breed the next year. Among the British butterflies adopting this strategy, often choosing garden sheds as a safe place, are the peacock and the brimstone, both familiar and welcome butterflies of early spring. Others breed and die, leaving the next generation to overwinter either as egg (the preferred strategy of members of the hairstreak family), chrysalis (small white and holly blue, among others), or caterpillar (by far the commonest overwintering form in the UK, and including such species as marbled white, ringlet and meadow brown). Whichever strategy is chosen, winter is tough. Because butterflies, like Marilyn Monroe's character Sugar Kane in the Billy Wilder film, like it hot.

To counter this, some, like the painted lady, have evolved to favour migration, and are prepared to travel thousands of kilometres so they can exploit the seasonal abundance of their chosen resources. This isn't just the natural dispersal around home breeding sites that other species – such as our swarming honey bees – might undergo, but 'true' migration – an intrinsic part of their life cycle, central to their survival strategy.

Every ten years or so there is an unusually large influx of painted ladies to the UK. These are 'painted-lady years'. It's thought these events might be associated with increased rainfall in the preceding months, with resulting abundance of food plants in the relevant areas. In these years painted ladies arrive in such numbers – and we're talking millions – that members of the public, not necessarily interested in or aware of butterflies in the normal course of events, find themselves noticing them. As well as giving excitement to butterfly-lovers, these influxes usually make the headlines of the national press, and do a lot to raise awareness of nature in general – and, counter to the usual representation of insects in the press, these are always 'feel-good' stories, and are therefore to be celebrated.

Once we've accepted that these butterflies are capable of such incredible journeys, and have got to grips with the underlying motivation for them, there remains the question of how. How does something as small as a butterfly fly so far? The distances involved are mind-boggling. That something so small, so apparently fragile, might travel 4,000 km as an individual and 15,000 km across a series of generations . . . well, it's the kind of situation for which the words 'gawping in disbelief' were coined. The mundane, pragmatic answer is that painted ladies (and butterflies and moths like them undertaking similar but shorter journeys in other parts of the world) are strong flyers, endowed with perseverance and an unquenchable evolutionary imperative. We can gawp all we like, but these creatures don't make such journeys on a whim. Nor do they do it to get into human record books. They do it because it offers their species the best chance of survival. If that means flying 2,500 km at a height of 500 metres then so be it. Last one to Morocco's a goner.

We love records. Biggest, smallest, heaviest, oldest, longest. They give us a sense of the scale of things. When you know that the smallest insect is less than half a millimetre long and the longest is 35.5 cm,* it helps you place it – smaller than dogs, bigger than amoeba. And so it is with journeys. We marvel at the painted lady's journey because it's the longest butterfly migration, outstripping

* It's a stick insect, which ever so slightly takes the gloss off it. Being long is what they do. That's measuring by body length. If you count specimens with their legs stretched out, the longest, *Phryganistria chinensis*, is 62.4 cm long.

even the famous migration of the monarch butterfly, which only manages a piffling 7,000-kilometre round trip.

The migration of the monarch – from southern Canada to Mexico and back – has been more thoroughly researched than the painted lady's, for the simple reason that it's been known about for longer. And while we're not sure exactly how migrating butterflies navigate, hypotheses abound.

The newly hatched butterfly has a brain the size of a pinhead, and no guide to show it the way, merely a nameless instinct of which we have scant understanding. We do know that to a certain extent they use the position of the sun for orientation. And because they have the ability to detect polarised light, it doesn't matter whether the sun is shining or not. Then there's the possibility of genetic memory – something implanted in their DNA that instructs them, 'Go south.' More pragmatic, and easier for us to conceptualise, is the use of landscape. Mountains, coastlines and even roads are all landmarks clearly visible from the air – again, the genetic memory might be more detailed: 'Go south until you see the really big mountain'. And finally in this bubbling pot of hypotheses, there's the possibility of geo-magnetism – experiments have shown that the Australian bogong moth uses Earth's magnetic field to find its way around, so there appears to be at the very least a working chance that this ability is replicated in other species.

As for how they know when to move, this is also an area fraught with mystery. To stay ahead of the game, they need to anticipate what's going to happen – a drop in temperature, the death of their favoured food plant, the arrival of potential predators – rather than react to things happening to them at that moment. Somehow, they know it's time to leave.

One of the difficulties encountered in monitoring these journeys is fairly obvious: insects are small. Sophisticated radar is excellent at detecting larger numbers of insects as they pass by, and can even identify insects down to species level, but things quickly become difficult when you try to attach transmitters to individual insects. While technology is advancing all the time, finding a way to tag small insects without adversely affecting their flight capabilities is a huge challenge. The '5 per cent rule' – anything weighing less than 5 per cent of the insect's body weight is assumed not to affect its performance – limits the size of such devices, and the difficulty of developing transmitters of a suitable size that can send a signal further than a few metres means that tracking individuals has so far proved an almost insurmountable challenge. The recent development of a tiny tracker specifically for monarch butterflies points the way forward. Weighing just 62 mg, and measuring 8 × 8 × 2.6 mm, it attaches to the butterfly's back and records temperature and light intensity.[2]

Human inventiveness moves forward, and bit by bit we uncover new information. No doubt, before long, we shall know enough about these journeys to fill in some gaps. And this knowledge will most likely open up new areas of research, pose more questions for the next generation of scientists and researchers, and will serve only to show us how little we understand of the creatures with whom we share a planet.

And that is exactly how it should be.

Of all insects, butterflies are probably the ones we like the most. They dazzle us with their colouration, large wings and endearing flutteriness, but are also intrinsically unthreatening. They don't sting or swarm, they don't buzz annoyingly around our heads, and they don't land on our food, eat it and then chuck it back up. Primary schools don't close when a butterfly pupa is found on the premises, and tabloids don't froth at the mouth at the thought of an influx of HORDES OF POISONOUS BUTTERFLIES.

Butterflies – despite the protestations of gardeners whose crops are ravaged on an annual basis by cabbage whites – are fine with us.

Moths, though. Poor things. They get such a bad press.

Strange, the psychology of it. The two are closely related and very similar. So much so that distinguishing one from the other often requires close examination. The usual criteria are the antennae – butterflies' antennae have little clubs on the end, while those on moths tend towards the fine and feathery (sometimes in an extremely attractive and endearing way). The way they hold their wings also differs, butterflies opting to hold them stiffly and vertically above the body, while moths opt for the more laid-back horizontal fold. There are exceptions, naturally, as there are with the guidelines that butterflies are colourful and moths drab – as anyone who has ever had a scarlet tiger moth in their garden will testify.

The reputation of moths in a population largely ignorant of their many charms rests predominantly – in the UK, at least – with two species (out of about 2,500, by the way – never has the reputation of so many been sullied by the behaviour of so few). They are the dreaded clothes moths (the common clothes moth and the case-bearing clothes moth), and it is reasonable to say that they're among the most cursed and reviled of insects.

Our perceptions are similarly coloured by our observation of their behaviours in flight. Butterflies are regarded as benevolent flitters and sippers, skittering their winsome way among the flowers and nectar-producing plants of the idyllic English country garden. Moths flail uselessly around the light bulb in the middle of the room, apparently under the deluded impression that it's the moon.

It all goes to show how little we know.

Part of this is down to when we're most likely to encounter the respective species. Butterflies (generally) are diurnal; moths (generally) nocturnal. Butterflies brightly coloured, moths drab. Butterflies stay outside, where they belong;* moths seem intent on bungling in through that open window in order to execute the aforementioned useless flapping light-bulb worship dance. If you did a straw poll, most people would prefer butterflies.

This is a shame, because for all sorts of reasons moths are no less worthy of our attention and admiration, and there is far more that unites them than separates them.

The main distinguishing feature of butterflies and moths can be gleaned (as so often) from the Greek words that make up their scientific name: Lepidoptera. I think we all know by now what *pteron* means. The other bit, *lepidos,* means 'scale', and it is by their scales, rather than their veins, that we shall know them.

These scales – thin plates, loosely attached to the body surface in individual sockets – are extremely small. Overlapping each other like roof tiles, they cover the whole body, legs and wings included. They come away easily. If you touch a butterfly or moth wing, they come off on your finger – a light dusting of powder.

* Except when they decide to overwinter in the folds of the curtains in the spare room.

Take away enough of them and you expose the wing membrane underneath. This loose connection is a form of protection – if the insect is caught in a spider's web, they're more likely to be able to escape, leaving the scales behind.

Looking at the varied functions of these scales, it's easy to see how butterflies and moths came to adopt them so widely. They insulate. They help with thermoregulation. They form a layer of trapped air next to the body, providing extra lift when the insect fancies a bit of a glide. But most obviously and visibly, they are responsible for the colours and patterns that, depending on the species, make us say, 'Ooh, look how pretty', or 'Meh, drab little thing, isn't it?'

While there is significant variation in the individual structure of the scales, they are generally constructed to the same plan: an envelope-like arrangement with two plates. The lower plate is smooth, but the upper plate is ridged, with intricate arrangements of crossribs and variations in the ridging patterns. These elaborate microstructures provide structural colouring – interfering with the light as it hits the wing – and add great variety to the insect's colour scheme. Sometimes this manifests itself as iridescence – the metallic sheen caused by light reflecting off the surface, and visible only when seen from certain angles.

Combining structural and pigmented colours, butterflies (and some moths – let's grant them that) achieve a wealth and variety of colour and patterning that is the envy of fashion designers the world over. Add to that the ultraviolet part of the spectrum, which not only helps them detect nectar guides on some plants, but also plays a role in mating communication, and it's clear that one way or another, colour is an important part of lepidopteran lifestyles.[3]

The uses of these colour schemes vary. For those gaudy butterflies the bright colours might be used to attract a mate. Or – as with the beetles already mentioned, and many other species – they might act as a warning: 'Do not eat me. I am disgusting and will probably kill you.' And the same principles of mimicry also apply: 'Do not eat me. I look like something disgusting that will probably kill you, but how can you be sure?'

Some of the more dramatic manifestations of bold patterning can be seen in the owl butterflies of Central and South America. They get their name from the spots on their wings, which resemble, to an almost creepy extent, an owl's eyes. Look at them long enough and you'll be convinced they're following you round the forest.

Opinion is divided as to the purpose of these eye spots. The obvious explanation is that they're there to intimidate potential predators. The eye spots stare the predator unblinking in the face, making it think it's up against something much bigger and scarier. Some even have glistening centres, and it is these that seem to have the most deterrent effect. Others have the eye spots on the hindwing and reveal them only when in immediate danger, giving the interloper a flash when they get close.

The other hypothesis is that these markings distract predators, encouraging them to attack non-vital parts of the body, like wingtips, rather than make a lunge for the essential organs. One species, the squinting bush brown butterfly, has a series of eye spots on the edge of the wing. The size of the spots varies, as different generations are born, according to the time of year. Small in the cool, dry season when they want to stay out of sight, and larger when the rainy season comes and they're out and about and wanting to feed and breed without ending up as a tasty snack.[4]

While butterflies often opt for gaudiness, moths tend towards the unassuming. In many cases, camouflage is the purpose, the intricacy of their different greys and browns often rendering them indistinguishable from the bit of bark they're sitting on. And their nocturnality also plays its part – there's not much point in developing a bright and varied colour scheme if nothing's going to be able to see it.

The wings of butterflies and moths send a message to the world, but their role as billboards is secondary. They're still wings, and they're still aerodynamically viable – their primary purpose is for flight.

Tracing the family history of butterflies and moths is hindered by an exceptionally sparse fossil record. The conditions required for fossilisation – broadly: burial, pressure, time – have not led to extensive examples of Lepidoptera fossils. On the face of it, this might seem illogical. Loads of other insects get fossilised – why should butterflies and moths miss out? One hypothesis is that because their wings are so good at repelling water, they're more likely to float and therefore more difficult to submerge. Whatever the reason, evidence is scant, so piecing together their history, and therefore ascertaining their precise place in the story of the development of flight, is difficult.

Butterflies and moths evolved shortly after flowering plants, taking advantage of the nutritional opportunities offered by their sweet, sweet nectar. They in turn served the plants by collecting and spreading their pollen, and so both groups grew and spread together, each servicing the needs of the other. This happened, so we gather, about 130 million years ago, the age of the oldest lepidopteran fossil, and by about 45 million years ago there had evolved animals recognisable as species alive today.

All this sounds plausible enough: a logical explanation for the diversification of both plants and Lepidoptera, and for their modern abundance and diversity. It's been the accepted position for quite some time.

Awkward, then, that in 2018 fossil evidence was found that pushed back the date of Lepidoptera evolution by about 70 million years, placing the first Lepidoptera fossils approximately 50 million years after the first Diptera, 20 million years after the first Hymenoptera, and roughly on the same page as the first true dragonflies.[5] These fossils are tiny – no bigger than a speck of dust – and were discovered accidentally while core drilling for pollen fossils in Triassic/Jurassic sediments in north Germany. Analysis of the approximately seventy scales showed them to have the distinctive structure of scales sported by modern Lepidoptera, and consistent with insects that had tubular sucking mouth parts. The significance of these minuscule things is that they pre-date the evolution of flowering plants, thus casting doubt on the idea that Lepidoptera co-evolved alongside them, and necessitating a rethink about the whole history of butterflies, and indeed of flowering plants.

We get used to one way of thinking, based on the available knowledge, and then a new discovery comes along that throws everything awry. One day there will be another, and then another, each one challenging the status quo. And so it continues, certainty as elusive as a shrinking bar of soap in the bathtub.

What we do know is that nowadays there are about 175,000 described species of Lepidoptera, of which just over 10 per cent are butterflies. As with other insect orders, that number probably represents a fraction of the true amount – another indication of the extent to which insects rule the world when it comes to diversity and abundance.

I have two favourite springtime hobbies: the first is trying to photograph a holly blue butterfly, and the second is failing to photograph a holly blue butterfly. These are inclusive hobbies, available to anyone with access to some sunshine, a camera and the slow reactions of a human being. The holly blue butterfly is small and flitty, silverish in flight despite the name. Its appearance in April is always a celebratory moment – one in a sequence of markers of spring's progress: crocus, blackthorn, lesser celandine, chiffchaff, sand martin, holly blue. Their colouring and delicacy make them a favourite of many butterfly fans, but there's no denying that they're flighty little devils, their scattery progress difficult to track. And when they do settle down (often on the holly that gives them their name) they like to spring back up again at the mere hint of the appearance of a camera lens. Hence my hobbies.

Other butterflies are easier to photograph. The peacock, large and comparatively easy to track in flight, also has the pleasing habit of stopping for a nice long bask in the sun, enabling even the most inexperienced and inept of insect photographers to get a good shot of its beguiling wing pattern – black-fringed maroon with dabs of cream and iridescent blue-purple 'eyes'. Summon a brief round of applause, too, for the red admiral, the common, large and showy garden butterfly, which is often quite happy to sit feeding on a mushy apple while you snap away to your heart's content.

Whatever the size of butterfly – and they range from the 1.5-cm wingspan of the western pygmy blue to the astonishing 30 cm of the Queen Alexandra's birdwing – their wings are large compared with their body. While this enables them to use their wings for display as already advertised, it has an aerodynamic effect – and not, on the face of it, a beneficial one.

The size of their wings means the ratio of body weight to wing surface area – wing loading – is low. And while this does enable them to generate plenty of lift, it's not a recipe for speedy, efficient flight. Butterflies, in general, flap their wings more slowly than moths, but the exact details of how they generate their flight forces have not always been clear. What they are good at is taking off. That quick launch is a useful predator evasion strategy, but it's what happens in the immediate aftermath that has only recently been revealed (after a few decades of educated conjecture). The wing tips, it turns out, meet at the top of the upstroke, clap together, and force air out behind them. This, of course, is the same clap-and-fling technique as adopted by much smaller insects. The upstroke isn't a recovery stroke, but is put to good aerodynamic use, providing thrust to complement the downstroke's lift. As with the smaller insects, but even more important because of the size of the wings, flexibility is key. Rigid wings would be much less effective, and wouldn't enable the cupping action that helps maximise the amount of air caught between the wings and then expelled.[6]

The standard wing arrangement, in both butterflies and moths, is not dissimilar to the linking system used by the Hymenoptera. In moths this takes the form of bristles on the base of the hindwing, which link, Velcro-like, to hooks on the underside of the forewing. In butterflies the arrangement is looser – the wings overlap, with a lobe on the hindwing pressing up on the underside of the forewing.[7] As with the Hymenoptera, this makes one aerofoil out of two surfaces. The dependence on the larger forewing for flight is highlighted by experiments that showed butterflies can still fly when their hindwings are removed. This does result in some loss of mobility, but slow flight is better than no flight.

The size of their wings and relatively slow flapping rates – the swallowtail manages an impressively zen-like five strokes a second – might give the impression of butterflies as clumsy flyers, but this is erroneous. They lack the super-zippy flight mode of nimble things like hoverflies, but their erratic flight serves them well, even if it does make them the very devil to follow. A butterfly expert will be able to identify a butterfly from a glimpse of its flight (this is a skill, like many, that eludes me, except with the most obvious species at the closest of quarters). Some have fast, direct flight – impossible to keep up with, even when sprinting; others skitter around unpredictably. While it might be tempting to ascribe this to temperament, the clue to this behaviour lies in toxicity. Toxic butterflies tend to have smoother flight patterns than non-toxic ones, which suggests that the scattery nature of flight is at least in part a predator evasion tactic. Toxic butterflies can rely on their unpalatability to keep predators away; the non-toxic ones need a slightly more sophisticated strategy.

The range of lepidopteran flight styles is large, and at the other end of the spectrum from the lazy lollop of the swallowtail is the smooth whirring hover of the hummingbird hawk moth. It should be relatively easy to work out how it got its name. To have one in your garden is one of the true treats of an English summer. They announce their arrival with the low breathy hum of their wings, find a suitable flower – jasmine, say, or verbena – and go about their business. They are just about large enough for the uninitiated to mistake them for hummingbirds, and their appearance, hovering in front of a flower, angled proboscis dipping into a flower trumpet for nectar, backs up this impression. By contrast with butterflies, their wings are a blur, as you'd expect from an insect that needs to stay still in the air to gather their life-giving elixir. Whirr, hover, sip, whirr – as good a lifestyle as any.

While insects were exploiting the advantages of life in the air, other animal life was developing in equally diverse and awe-inducing ways, both in the water and on land. But one thing united it all: it remained steadfastly earthbound. Species came and went without exploring the idea that up, if not the only way, was at least a possibility. A lot of these life forms were too heavy to fly, or had no need of the ability. Many more of a suitable size managed perfectly well without making the extensive readjustments required. If nothing else, this is a marker of just how difficult the whole affair is, and how circumstances need to conspire for it to happen at all.

But eventually it did happen. The first vertebrates to take the leap – metaphorical and physical – into an aerial lifestyle appeared shortly after the end-Permian 'Great Dying'. And they count among their number some of the most spectacular flyers of all.

7

THE PTEROSAUR

It's one of those good walks, steep at the beginning and levelling off as you near the top. The wind whips in from the sea. Healthy, bracing. Ravens might accompany you part of the way, gronking as they tumble and soar and ride the wind in that playful way ravens have. A skylark might give its trilling song from high above, invisible in the clear blue sky.

Below you is Sandown, the beach curving gently away into the distance. Pier, arcades, crazy golf. There is a zoo, devoted to living animals; there is the Dinosaur Isle museum, devoted to dead ones. You reach the top of the hill, invigorated by the exercise, but not exhausted by it. Panoramic views across the island, your reward for the climb. The wind dies down a bit, and out comes the sun.

All, for the moment, is well.

Then a strange thing happens. You catch a glimpse of something coming towards you from the other side of the island. A dot at first, morphing into a moving shape, not quite recognisable as anything yet, but moving fast, flying with ease and economy. It's difficult to tell, against the patchwork backdrop of the island's fields and villages, what it is. Not a bird. Not a plane.

Inevitably, you think of Superman.

And then it gets close enough for you to attempt an identification, and you have to take a breath. For a surreal moment you think it's a huge bat, and your mind reels at the oddness of it. But it's not a huge bat, odd though that would be. It is in fact something far odder. Its wings are broad, body slender. But the main distinguishing feature, revealing itself as it comes closer and wheels away from you, is a huge crest on its head. Not the wispy adornment favoured by some birds, but an apparently solid expanse of bone or cartilage. And you realise that this is a pterosaur, and you appear to have travelled 127 million years back in time.

Bloody hell.

It isn't, of course. And you haven't. It's merely a product of your fevered imagination. Or, more accurately, mine. Because not far from here a bone was found, and since I first read about it, I've been on a flight of fancy.

It doesn't matter for the moment what the bone was – we'll come to that. The main thing is what it, and the thousands like it discovered around the world since 1784, represents: our only connection with the first flying vertebrates on this planet.

Until the pterosaurs came along, flight was the preserve of the small and chitinous. Pterosaurs changed the rules. You could be big and fly; you could have bones and fly; you could adapt existing limbs into wings, rather than growing them out of your body from scratch.

The development of flight in pterosaurs must have been quite the shock for the insects. For millions of years they'd owned the planet's airspace. Now there were actual monsters flying alongside them, chasing them, and often eating them.

What we know about pterosaurs comes only from those bones. With all the other flyers – insects, birds, bats – even though those original species have long been extinct, we can point to a living thing and say, 'Look – it was something like that.' But with pterosaurs all we have are the fossils.

These creatures exist largely in our imaginations. We can never know what it is like to encounter a pterosaur, to see it flap into view over the sea, to track its progress across the sky, to fix it in our binoculars and drink in every aspect of its pterosauriness.

But we can have a go.

'There are known knowns; there are things we know we know. We also know there are known unknowns; that is to say we know there are some things we do not know. But there are also unknown unknowns – the ones we don't know we don't know.'

Donald Rumsfeld was widely mocked for those forty-six words, but once you pick the bones out of it there is little to argue with. It's logical enough thinking, expressed inelegantly.

He might have been talking about the fossil record. And in the case of the pterosaurs, there are many unknowns. But there are a few knowns, so let's start with them.

Firstly, pterosaurs aren't dinosaurs. This might be dismaying to anyone who has – if they were even aware of them at all – lumped them together under the catch-all heading 'Olden Monster Things'. But life is full of disappointments.

They were related to dinosaurs. Exactly how close the relationship, and what their common ancestor was, isn't certain, but current thinking seems to be that they were close enough to be on the wedding invitation 'A' list, sitting at a table marked 'Archosaurs' with the other members of the family, crocodiles.

Like dinosaurs, they were reptilian; like dinosaurs, they lived from the late Triassic to the end of the Cretaceous (although the dinosaurs did have a few million years' head start); like dinosaurs, they exert a fascination unlike any living thing. But there was one thing they could do that dinosaurs couldn't.* They could fly. All of them. There were, as far as we know, no flightless pterosaurs.

That phrase: 'as far as we know'. We're going to need it a lot. Assume it applies from hereon in. Because one other thing we know about pterosaurs is this: we don't have nearly enough of them. The fossil record is frustratingly sparse. And this leads to hypothesis, conjecture and educated guesswork as pterosaur workers extrapolate as much information as they can from existing material and try to fill in the gaps.

For many people of a certain age, the word 'pterosaur' might cause some confusion, a raised hand, and a puzzled question: 'Is that the same as a pterodactyl?' At which point the pterosaur worker heaves a sigh and begins the explanation.

* With the exception, of course, of the dinosaurs that later came to be known as 'birds'.

The word *ptéro-dactyle* ('wing-finger') was coined by French scientist Georges Cuvier in 1809,* and was applied to the aforementioned 1784 fossil, found in the Solnhofen limestone in Bavaria and described originally by Cosimo Collini. Collini thought it was an aquatic animal, but both Cuvier and French-German zoologist Johann Hermann agreed that the elongated fourth finger, clearly visible in the spillikins jumble of angular bones, might support a membranous wing. Hermann reckoned it to be a flying mammal, and his drawing of it, featuring huge, rounded, billowing wings, is at least consistent with that view, even if it's not clear exactly how he expected it to fly with bed sheets attached to its arms. Another artistic representation in 1843, by Edward Newman, has the pterosaurs looking endearingly like winged possums, with big button eyes and long, toothy snouts.

Years passed, more fossils were discovered, and the nature of these animals gradually became clearer. But even though the order Pterosauria was officially named by Richard Owen in 1842, the name 'pterodactyl' seems to have endured in the public consciousness much more successfully. That original 'ptéro-dactyle' was renamed *Pterodactylus*, and those animals now take their place as one of about twenty pterosaur families, along with such linguistic delights as Rhamphorhynchidae, Anurognathidae, Azhdarchidae, Ctenochasmatidae and several others of a similar bent.

Which leads us to the thorny area of pterosaur names. The whole subject of pterosaur classification is confusing. Part of this confusion, for me at least, comes from the names. They vary in

* His paper originally contained the misprint 'petro-dactyle', which conjures up images of a completely different kind of animal.

type. Some are hybrids of the place the fossils were found, bastardised into 'scientific Latin/Greek'; others are directly translatable from the Latin (*Rhamphorhynchus* = 'beak snout'), which is great if you have a bit of Latin knocking about the place, but problematic if you don't. And because humans have never lived alongside them, there's a complete absence of the kind of vernacular or folk names given to other animals. While it's entirely understandable that scientists want to be specific about such things, it does leave the rest of us grappling with new and unfamiliar words and ever so slightly trailing in their wake. Even though you do eventually get the hang of it, it's nevertheless tempting to suggest a completely new naming system using recognisable English words. We've done it with birds, so why not with pterosaurs? If *Passer domesticus* can be a house sparrow, why can't *Dsungaripterο weii* be a crested broadwing? Or *Rhamphorhynchus longicaudus* a snaggle-toothed vanetail?

Until my brave new world of English names for pterosaurs kicks in, we have the scientific names. Some of them I have to say slowly several times before they sink in. And the involved history of our understanding of their evolution has only compounded that confusion. For a long time pterosaur research languished. But since the 1980s there has been a surge of discoveries from all over the world, and now we have more than 200 of them – the majority found and described in the last twenty years – with the promise of more to come. These new discoveries necessitated rejiggings, creation of new families and reassessment of exactly how the various genera were related to each other.

It's easy to assume, coming to the subject cold, that all pterosaurs lived at the same time. It's all a very long time ago, after all, and timespans get compressed all too easily into a vague era with the catch-all title 'prehistory'. But they were around, in one form or another, for over 160 million years. They lived all over the world, and in habitats ranging from forests to the open ocean. Species came and went, as is the way of things, and over that period there were clear developments in how and where they lived, as well as, most noticeably, their size.

Pterosaurs have conventionally been divided into two groups: rhamphorhynchoids and pterodactyloids. Broadly speaking, the rhamphorhynchoids lived earlier and were long-tailed and comparatively small, while the pterodactyloids lived later and were short-tailed and larger. The usual caveats about exceptions apply. And grouping together a ragtag assemblage of different families under the broad umbrella 'rhamphorhynchoids' has more to do with convenience than an intrinsic family relationship binding them together. There has been a move towards calling the rhamphorhynchoids 'non-pterodactyloids', but that way lies not only confusion but a veritable barrage of the word 'pterodactyloid', of which we've already had more than enough. So 'rhamphorhynchoids'* – here, at least – they remain.

Whatever group they belonged to, and consistently across the 160 million years of their existence, pterosaurs of all shapes and sizes were united by some basic anatomical features that set them apart from any other creature, living or dead.

Shall we? Let's.

* I'm aware that in making my point I have also used way more than the usual quota of the word 'rhamphorhynchoids' in this paragraph – a situation only compounded by this footnote.

If you were uncovering a pterosaur's wing bit by bit from the shoulder, you would find nothing particularly unusual until you got to the wrist. The upper arm – humerus – is fairly short and stout, and articulates with the forearm via a standard elbow joint that allows nearly 180 degrees of rotation. The forearm – two parallel bones, radius and ulna – is longer but not excessive. It reaches the wrist perhaps a bit further away than you might expect. The wrist itself is straightforward – fused into two bony blocks with limited flexibility.

This is where it starts to get odd. Sprouting out of the wrist is a bone found in no other animal. Ranging in size from little more than a nub in early pterosaurs to a thin rod in later animals, it's called the pteroid, and its function was to support a little expanse of wing membrane called the propatagium.[1]

The problem with the pteroid is that there are several ways you could attach it to the wrist, and as pterosaurs lacked the simple courtesy to be preserved in precise poses that accurately represent the orientation of their skeletons in life, the position remains unclear. Did it point forwards, into the wind, making the propatagium quite large? Or was its angle to the wing shallower? Might it even be deployed in different positions according to circumstance? And what difference might it have made to a pterosaur's aerodynamic performance?

The pteroid is one of those little things, like the dragonfly's pterostigmata, that might – or, alternatively, might not – have a disproportionate effect on an animal's flight capability. The prevailing opinion seems to be that it didn't point forward but that it was adjustable, allowing the pterosaur to deploy it as an aid to manoeuvrability if necessary.

But while such marginal features are of some interest – and this one has nagged away at me ever since I first read about it – the really strange thing about the pterosaur's wing is yet to come.

After the wrist, the fingers. One, two, three, each slightly longer than the last.

Then the fourth finger. It goes on. And on. And on. And on. A pterosaur's fourth finger is noticeably, weirdly, excessively long. A veritable überfinger. The kind of thing you'd use to get a ping-pong ball out from behind a radiator.

They did not use it for that though. They used it to support the brachiopatagium (this translates as 'arm membrane', which is just as well, because that is exactly what it is). This large expanse of membrane sweeps down from the tip of that fourth finger all the way to . . . well, where exactly? The usual caveats about soft tissue preservation apply to these membranes. Helpful examples are very rare. The fossilisation process never allows the membranes to be found exactly as they were when fully stretched out in life – there is always a certain amount of folding. This has led to considerable uncertainty about exactly where the membrane attached. Interpretations range from the ankle to the thigh and every point in between, with the most broadly favoured spot currently the ankle. There is also the likelihood that it will have varied from species to species.

You might think that with the wings fully stretched this would provide a massive expanse of membrane stretching directly from the finger to the ankle, giving an appearance something like Dracula's cape, but in reality the brachiopatagium would most likely have followed the contour of the arm bones before sweeping down to the ankle.[2]

There are two things arising from this. Firstly, pterosaurs' wingspans generally seem disproportionately large to anyone used to birds; and secondly, when they're not in flight, the wings don't fold tight against the body the way they do in birds. Instead, pterosaurs move quadrupedally, and the point of contact with the ground is the hand, with that fourth finger – with its attached wing membrane hanging down like a closed curtain – pointing straight up in the air.

This would have given them a posture, while walking, not dissimilar to a folding canvas picnic chair. Pick it up by the tail, give it a shake, and it might unfold with a click.

In the membrane itself there is more than immediately meets the eye. It would be easy to assume, as scientists did for many years, that this was a thin single sheet analogous to a bat's wing. But the crucial difference between the two is in the hand structure. A bat's wing membrane is supported from within by four splayed fingers. The pterosaur's membrane, hanging from that single, massively elongated finger, lacks that support – if it were as thin as a bat's, it would be too loose and flappy to be aerodynamically useful. What it needs is tension. This is provided partly by the structural strength of the wing bones, but also by the structure of the membrane itself. It has an inherent thickness, complexity and strength far more sophisticated than is first apparent. It takes a form not, in principle, unlike an insect's wing – a sandwich of three layers, each contributing to the membrane's strength and therefore playing an important role in a pterosaur's flight capabilities. On the underside, a thick system of veins; in the middle, a thin layer of muscle and connective tissues; on top, flexible fibres – actinofibres – less than a millimetre wide. These are concentrated more densely towards the wingtip and help give the wing strength and structure. The combination of those fibres and the sheet of muscle would have given the pterosaur the kind of control over the wing provided for a bat by its fingers.

Completing the survey of pterosaur membranes, we come to the uropatagium, a little thin stretch of the stuff between the legs. Again, the extent of this varied according to species, from full skirting to baggy trousers. Very rarely preserved, this would probably have been thin and fragile, contributing a little extra lift, if that.

The upshot of this combination of long limbs and strong membranes was that pterosaurs were, for the most part, excellent flyers. They came in all sorts of shapes and sizes, of course, each adapted to different flying styles, but the general body plan gave them strength and flexibility.

It's one thing being able to propel yourself through the air, but quite another to get up there in the first place, and there remains the question of how pterosaurs took off. Did they launch themselves after a long run-up? Throw themselves off cliffs? Glide down from treetops? Perhaps they flung each other into the air with a primitive trebuchet.

Yet again, recent analysis seems to have provided the answer.[3] Pterosaurs most likely launched from a sitting position on the ground. First a crouch back, then the spring, propelled mostly by the powerful forelimbs. Having relatively flimsy hindlimbs was no hindrance to their take-off abilities. In fact it was an advantage – the launch muscles and flight muscles are one and the same, saving valuable weight. Compare this with birds, which mostly use their leg muscles to launch. Once the bird's in the air, the legs are dead weight. An inefficient distribution of resources, birds. What were you thinking? For pterosaurs, that first spring needed to be powerful enough to clear the ground for the first wingbeat, and then they were away.

The membranes are only part of the story. Pterosaurs' bodies were, to a greater or lesser extent, adorned with feather-like structures called pycnofibres. To describe them as 'feather-like' might seem to be taking a liberty, given that these were single, shaftless filaments. We'll examine what is or isn't a feather when it comes to exploring the evolution of birds, but the similarity of pycnofibres to the most basic of feathers has led to conjecture that such things first arose in the ancestors of dinosaurs and pterosaurs, evolving to full and complex feathers in the former and grinding into a dead end in the latter. What purpose pycnofibres might have had is difficult to evaluate – their sketchy presence on many pterosaurs would seem to indicate that their value as insulation was limited, but in a few species they seem to have amounted to an almost full-body covering of thin fur.

The final bodily feature distinguishing pterosaurs was their head crests. Bony protrusions, often enhanced by soft tissue, in some cases they were quite spectacular, and in others just plain odd. It was once suggested that these crests were used in some way as rudders to aid flying, but the consensus is now overwhelmingly that they were used for sexual selection – the more extravagant the crest, the more likely the male was to attract a female. And while you might assume that sporting a big lump on your head might hinder your progress through the air, experiments have shown that in some cases these crests might not have been as detrimental to flight capabilities as you might imagine.

While they were united by these body features, there was considerable diversity in the shapes and sizes of pterosaurs from their emergence in the late Triassic to their (spoiler alert) demise at the end of the Cretaceous.

Let's meet a few of them.

The first pterosaur, until someone finds something earlier, appears in the fossil record about 228 million years ago, in the Preone valley in the Italian Alps: *Preondactylus buffarinii* ('fingered thing of the Preone valley, named after Buffarini'). This is the late Triassic. Griffinflies have left the stage, flies and bees are just about getting going, and butterflies are mere twinkles in evolution's eye.

We go there, somehow. We ignore the inconvenient impossibility of time travel, and the awkward truth that the levels of oxygen and carbon dioxide in the atmosphere would not be conducive to supporting human life. We also ignore the strong probability that within ten minutes of landing we would be torn apart – or simply eaten whole – by a dinosaur, and that even if we did somehow manage to survive and find shelter, we'd almost certainly succumb to a disease caused by the very different set of microbes that prevailed in the late Triassic. And – if we're doing this – let's say we manage to avoid the dinosaurs and the microbes and find ourselves within viewing distance of a not overwhelmingly gigantic pterosaur, it's a toss-up whether it would view us as a tasty morsel, a looming threat or a benevolent visitor to be allowed to go about our business as we please (hint: it's probably not the last one).

Mercifully for this silly experiment, *Preondactylus* is in fact one of the smaller specimens, as they tended to be around that time. About the size of a woodpecker, with a wingspan of just under half a metre. Reproductions of *Preondactylus,* based on the fossil evidence of a few bones, depict it with a long, toothy snout, bright eyes, and a long, straight tail. It would, so it's reckoned from analysis of its single-cusped teeth, have eaten insects. Or fish. Or both. Maybe.

The vague details of its life are, in a way, irrelevant. It holds the same fascination as *Rhyniognatha hirsti* or *Delitzschala bitterfeldensis*: it was (or might have been) the first. The uncertainty is part of the fun. And that's made all the more alluring by the fact that we don't know what came before it. There is a distinct lack of what you might call proto-pterosaur fossils. But extrapolating from certain features of early pterosaur anatomy – clawed feet that could help them climb, and their probable ability to splay their hind legs, to name but two – the likelihood seems to be that although later pterosaurs launched from the ground, the origins of their flight were in the trees. This is a debate we'll revisit in the discussion of birds and their route to flightiness, so for now we shelve it with an enigmatic smile and a tap to the side of the nose.

These early pterosaurs were already competent flyers, which would indicate that the transition from non-flying species to flying species was relatively fast. Perhaps understandably, though, with short wings and legs, *Preondactylus* showed characteristics consistent with squirrel-like competence both in trees and on the ground.

Life being life, the *Preondactylus* and other early pterosaurs went extinct – everything does, sooner or later – and others took their place. Comparatively little is known about the extinction event that marked the end of the Triassic and the beginning of the Jurassic 201 million years ago, but while it did for plenty of marine reptiles – as well as land-dwellers such as thecodonts – dinosaurs and pterosaurs emerged from it comparatively unscathed. And the Jurassic saw not just a growth in the number of pterosaur species, but diversification in body type and behaviour, too.

Take *Rhamphorhynchus muensteri* as an archetypal Jurassic pterosaur. These narrow-winged, long-tailed animals are abundant in the fossil record of the Solnhofen limestone, many specimens featuring the coveted soft-tissue preservation that helps build a more complete picture of pterosaur anatomy. The size range of the fossils is wide – the smallest, at 29 cm, has the wingspan of a swallow, while the largest, 1.8 metres wide, is closer to an eagle's – leading researchers initially to identify several separate species within the genus. But now it's thought that the size variations mostly represent different stages of growth within the one species, from juvenile (endearingly known as 'flaplings'*) to fully grown adult. Although *Rhamphorhynchus* is just a representative sample, there's strong evidence that all pterosaur flaplings were strong flyers from a very early age – even potentially flying almost immediately after hatching. Their body plans differed from those of adults, developing relatively quickly before a growth slowdown after a couple of years. The combination of low body weight, strong bones and high-aspect-ratio wings would have meant that flaplings were able

* A term proposed by David Unwin in his 2005 book *The Pterosaurs from Deep Time.*

to combine flying and gliding, living lives independent of their parents and occupying different habitats and ecological niches.

Another pivotal pterosaur species of the Jurassic was *Darwinopterus modularis*. Hailed as a groundbreaking fossil on its discovery in 2009,[4] it's a sort of chimera, combining characteristics of early basal pterosaurs with those of later, more derived ones – with the long head and neck of the later pterodactyloids, but everything else, including the trademark long tail, following the template of the rhamphorhynchoids. A flatpack self-assembly pterosaur with the parts mixed up. Its discovery also rattled the evolutionary cage, offering an argument for modular evolution – the idea that evolution happens to some bits of the body without affecting others – in pterosaurs.

Pterosaurs sometimes seem as if they're the result of the over-fertile imagination of an artist with unlimited access to medieval dragon illustrations and a copious supply of mind-altering drugs. *Darwinopterus*, perhaps as much as any, embodies this impression. Looking at paleoartists' renditions of these gawky, angular creatures, it's not a stretch to imagine fire emerging from their mouths as they flap through the gingkos and conifers of Jurassic forests.

But of all the pterosaurs of the Jurassic, the ones that draw my eye are the Anurognathidae ('tailless jaw'). Small for pterosaurs, and among the first with short tails, these were endearing, frog-mouthed, huge-eyed insectivores with jaws at least as wide as they were long. Those massive eyes are a sure indicator of cuteness, but also, more scientifically, that anurognathids were likely to be crepuscular or nocturnal. And all the evidence points to their being forest-dwellers.[5]

Most of all, though, their short, broad and powerful wings seem well suited to extremely agile and manoeuvrable flight – analogous to, for example, swallows, swifts and bats today with their ability to pluck insects out of mid-air. Quite the shock for the insects of the air, which had lived for millions of years fearing only consumption by their own kind. Just one more bloody thing for them to worry about.

While anurognathids developed the necessary agility for aerial hunting, other pterosaurs looked downwards, relying on less mobile targets. One such was the stumpy-legged, long-necked *Pterodaustro*, which hung around shallow lakes in Argentina (and possibly other places, for all we know) about 105 million years ago. Its large, strongly upturned jaw housed the slenderest teeth known in any animal – over a thousand of them, densely packed and bristle-like, reminiscent of the finest of fine-toothed combs – a similar arrangement to a whale's baleen. The teeth, allied to the presence of gizzard stones, or gastroliths, in its stomach – consumed to help with the breaking down of food – would seem to indicate that *Pterodaustro* was a filter feeder, rather like a flamingo, scooping up mouthfuls of water and straining it through the fine mesh of its teeth. To judge by its size, shape and weight, *Pterodaustro* flew only with difficulty, limited to taking off at a low angle with energetic flapping to lift it ponderously into the air.[6]

Pterodaustro's crest was shallow and discreet, and anurognathids were notable as the only completely crestless pterosaur family. At the other end of the cresty spectrum were the tapejarids, the group containing my Isle of Wight fever dream, *Wightia declivirostris*. They are notable not just because they sported some of the most extravagant crests of all pterosaurs – the kind of thing you might see at Royal Ascot – but also because they're thought to be the only pterosaurs that were predominantly herbivorous. Parrot-beaked and probably only moderate flyers, their members nevertheless survived for over 50 million years between them, spanning a period of great change for pterosaurs. One of them, *Tupandactylus imperator*, with a wingspan of four metres, even hinted at the future for the pterosaurs of the late Cretaceous: gigantism.

Pterosaurs got bigger. Which isn't to say that the small ones died out completely. But as they moved through the Jurassic and into the Cretaceous (145 to 66 million years ago), the wingspans of some burst out of the 1–2 metre bracket, and cracked on towards 4 and 5 metres, and eventually to sizes that beggar belief. For context, the largest wingspan of any living bird is the comparatively puny 3.5 metres of the wandering albatross.

If you're looking for a Cretaceous pterosaur comparable to the albatrosses, your eye would be drawn to *Pteranodon* ('toothless wing'). Known from more than a thousand specimens, and with wingspans ranging from 4 to 8 metres, this is a pterosaur often represented in images of the late Cretaceous. Its signature boomerang crest and long upturned beak (which gives the impression that it's in a permanent state of wry amusement) mean that along with *Pterodactylus* it's the creature most likely to spring to mind at the mention of the word 'pterosaur'. Those 1,000-plus specimens are mostly very fragmentary, but between them they cover most of *Pteranodon*'s body, so it's been possible to build a reliable picture of what they looked like and how they behaved.

The vast majority of these fossils are from an inland sea in what is now Kansas. In the late Cretaceous this would have been more than a hundred kilometres from the coastline, making *Pteranodon* a strongly ocean-going animal, using those vast wings to soar effortlessly over the waves, and coming to land only to breed.[7]

Also ranging across the ocean were several species of nyctosaurid – closely related to *Pteranodon*, and with a similar body plan, but significantly smaller. The prevalence of just these two families of pelagic pterosaurs in the late Cretaceous – both sizeable and comparable to the soaring birds we have now – does raise the question of what else might have been keeping them company. There would have been plenty of scope for smaller pterosaurs to fill the available niche, much like puffins and petrels and other associated bird species do today. So far, such fossils are notable for their absence. *Pteranodon* would have presented an imposing figure as it soared over the ocean waves, but even its 8-metre wingspan was outstripped by two pterosaurs vying for the title of 'Biggest Flyer Ever'.

The name *Quetzalcoatlus northropi* (Quetzalcoatl was an Aztec winged serpent god; John Knudsen Northrop was the founder of an American aircraft manufacturer) might not trip off the tongue, but it's likely, if you have any interest in the subject, you will have seen an illustration or model of it. It's another of the glamour pterosaurs, and all because of its one simple trick: it's absolutely massive. And good luck to it. My gaze was certainly drawn to it, and I remain susceptible to open-mouthed wonder at its sheer hugeness.

Take a giraffe. Slim the neck down a bit, swap out the nose for a long, dagger-like beak, and transform the front legs into a pair of massive wings. You now have – more or less, and conveniently ignoring internal anatomy and other such arcana – *Quetzalcoatlus northropi*. And while we might be used to the thought of a giraffe as a benevolent animal, meditatively chewing on acacia leaves, turning its inquisitive gaze gently towards the approaching camera operator, and occasionally breaking out into an unthreatening trot, the image of *Quetzalcoatlus* is of a different order. This was a very big animal indeed, and it flew. Its wingspan was, it's reckoned, at least 10 metres. That's half a cricket pitch, six times longer than my kitchen table, or (for those who like their measurements in more traditional units) one-fifth of Nelson's Column.

How then, the inquisitive reader will already be asking, did the darn thing get off the ground?

Good question.

Working out how big *Quetzalcoatlus* was, and how much it weighed, has been a source of contention ever since the first fossil was found in 1971. The problem being not just that said bone – part of a wing – represented a small fraction of the animal's body, but also that the specific anatomy of these creatures has always been far from clear. Crucially, the composition of the bones has caused debate and reassessments over many years, and only recently has it been agreed that while their bone walls were extremely thin and the bones filled with air, this didn't mean they attained the miraculously light weights proposed a few decades ago. Estimates of its weight have ranged from 85 kg to 540 kg, which, if nothing else, tells you how difficult it is to be categorical about such things. The latest estimates seem to have found a happy medium, settling on something between 220 and 250 kg – heavy enough to mean that they were operating close to the limits of aerial locomotion.[8]

Would you want a *Quetzalcoatlus* at the bottom of your garden? Probably not. Would you, on the other hand, want to see one sweeping over your local high street in all its majestic angularity, eliciting awestruck comments from passers-by? Chalk that one up as an emphatic 'yes' from me.

The other contender in the 'fuck me, did you see that?' stakes is *Hatzegopteryx thambema* ('terror wing from the Hatzeg basin'). Estimates of its wingspan are in the same range as for *Quetzalcoatlus*, but differentiating it from its cousin – they are both members of the group known as Azhdarchidae (from the Persian word for a mythical dragon-like being) – is its gigantic skull, reckoned to be up to 2.5 metres long. With its heavily muscled, comparatively short neck, and bones that were spongy rather than hollow inside, it would have been smaller than *Quetzalcoatlus*, but also heavier. But in both cases, their extreme size would have enabled them to roam across huge distances, some even suggesting that these and other massive pterosaurs might have ranged globally.

Estimates of the largest potential flying size vary, but it seems that animals such as *Quetzalcoatlus* and *Hatzegopteryx* were pushing the limits. Even at 10 metres, the margins would have been fairly slim. Beyond about 12 metres, it's likely they would have been too heavy to get off the ground and their bones too fragile to bear the strain.[9]

It's worth saying again at this point, given the preponderance of arse-covering in the above passage, that the fossil record precludes certainty on many of these issues. Several of these species are known only from a handful of findings, and some from just one. Each new discovery sheds fresh light on an area of palaeontology that exists largely in the shade. Pterosaur workers, armed with scant information, are not only expert in extrapolating from tiny amounts of information – they're also skilled bet-hedgers. And quite right too.

I walk down the hill, past the golf and the zoo. I pick up an ice-cream, making sure to shield it from a curious herring gull. Into the Dinosaur Isle museum, where dramatic models of long-extinct animals rub shoulders with cabinets containing real fossils. Education balanced with entertainment. Children fill in activity forms; adults wait patiently. I read everything, look at everything, submit myself to the time machine.

In one corner, model pterosaur heads stick out of the wall, one above the other. They are mostly ignored in favour of a nearby juvenile T. rex. Raise your sights and there are models suspended from the ceiling, flying harmlessly overhead, easy to miss unless you look up.

I just want to see one. That's all. One of them, for five minutes. OK, maybe ten. A living pterosaur, in all its cresty splendour, sitting on the ground a kilometre away, then springing to the air and flying over my head with slow, lazy wingbeats and away across the sea.

It's tempting to imagine a world in which the pterosaurs survived beyond the K–Pg extinction. Conjecture is great fun. We can't stop ourselves. Would they have made it this far? Or would they have been outcompeted by the birds? Might a visit to a wildlife reserve nowadays take in the descendants of *Pteranodon, Nyctosaurus, Quetzalcoatlus* and the like? We would doubtless have come up with different names for them.

For the 160 million years or so of their existence, they changed what it meant to be a flyer. But even as they were diversifying, growing, adapting to different habitats and exploring the limits of aerodynamic possibility, their cousins – or some of them at least – were developing, too. And soon they would join them in the air.

8

THE ARCHAEOPTERYX

I was a tiresome child. A factoid-obsessed bore, never happier than when interrupting a grown-up conversation to observe – completely out of context – that the capital of Mongolia was Ulaanbaatar or that the country with the lowest average annual rainfall was Libya. I knew the height of Mount Everest (8,849 m), the distance from Paris to Madrid (1,199 km), Donald Bradman's batting average (99.94). I memorised π to 100 decimal places (I can still do the first thirty-five: 3.14159265358979323846264338327950288) without ever considering either that this was an entirely useless skill or that π might have significance beyond enabling eleven-year-olds to show off their memory to a room of bemused adults. Little factoids, devoured and squirrelled away for no better reason than the devouring and squirrelling, they took the place of real learning. It was – I was – unbelievably irritating.

But there was one fact that has proved, if not useful, at least a springboard for curiosity later in life.

'The first bird was *Archaeopteryx lithographica*. It lived 150 million years ago and the first fossil was found in 1861 by Hermann von Meyer.'

I didn't question it. Nor did I feel the need to explore further. Knowing the first bird was enough for me. The second, third and fourth birds, those also-rans of ornithological prehistory, held no interest. Losers.

Only much, much later did I learn that, as so often in life, the whole story was more complicated, less clear-cut, and far more interesting.

In 1857 Hermann von Meyer contemplated a jumble of bones that had been found two years earlier in the Solnhofen limestone in the Altmühl Valley in Germany. He identified it as a pterosaur – one of many found in the limestone of the area that provides superb conditions for fossilisation. He named it *Pterodactylus crassipes* ('thick-footed winged-lizard').

Remember this. We'll need it later.

Whether he was excited by the finding – the Solnhofen limestone yielded a lot of pterosaur fossils – was rendered irrelevant within four years by the discovery of first a fossil feather, and then a skeleton. Once again it was the Solnhofen limestone that gave up its treasure, but this time the fossils pointed to something completely new.

The feather is delicately preserved in two counterslabs. It looks, on first inspection, very much like the kind of thing a careless pigeon might leave behind on your local high street.

The skeleton, found a few kilometres away, is of a different order. It lacks a head, one foot and some fingers. The lower body parts are relatively complete although broken up and scattered. But it's the wings that catch the eye. Fanned out, with clear impressions of their feathers surrounding folded bones, they give the skeleton the appearance of a gnarled angel.

Displaying a mixture of avian and reptilian features, it was later described by Charles Darwin as possessing 'a long lizard-like tail, bearing a pair of feathers on each joint, and with its wings furnished with two free claws'.

Taken together, the feather and the skeleton ranked – and still rank – among the most significant fossil findings of all.

Here's what you need to do if you're a living thing looking to become a fossil.

First, you need to die. This is an awkward but unavoidable first step. If you can see your way towards doing so somewhere near water, that's an advantage. And if you can do it in a place where there are shifting sediments, so much the better. Basically, you need to be submerged, and sharpish. Oxygen is your enemy. The application of pressure, for mineralisation of the relevant body parts, is the gold standard. The longer you remain unsubmerged, the more likely it is that your soft bits – flesh, membrane and so on – will be scavenged or decompose. In fact, if you really have your heart set on soft-bit preservation, you might want to lower your sights – this outcome is very far from the norm. Probably best just to aim for the clear preservation of a single bone, say, or, if you're a member of the chitinous community, a wing or two or a bit of exoskeleton. Hard bits fossilise far more often than soft bits.

If you're really lucky, there will be a volcanic eruption nearby shortly after your sad demise – the rapid accumulation of ash covering your body before it has had a chance to decompose. This might even be the thing that kills you, in which case you will be preserved in death's throes, your agony immortalised.

Then you wait. If you're lucky, anywhere between a few thousand and a few million years later you will be uncovered by someone who knows what they're doing, and they will take great care to treat your remains with the respect they deserve. If you're really lucky, you might end up in a museum. It all rather depends on how important humans think you are. If you're 'just another trilobite' then the prognosis is, frankly, poor. But if you were an individual with the potential to shed light on exactly what the hell was going on at the time you lived, then you can expect the full VIP treatment. You will be examined, pored over, discussed. Papers will be written about you. Counter-arguments will be made, followed by counter-counter-arguments, rebuttals and refutations. You will be branded 'significant', 'important', 'groundbreaking'. Oh, the fun they will have.

Or you might end up in a petrol tank, your contribution to the world in death no more than to keep a Ford Mondeo ticking over at the traffic lights before your remains are burned and the fumes dispersed to the four winds to become once more part of the space dust of which we are all made.

The likelihood, though, is more mundane. You will probably stay there for ever, your life forgotten, never touching the consciousness of those who come after you. Rest in peace.

Timing is important. Just two years before the discovery of the Solnhofen *Archaeopteryx*, Darwin had published *On the Origin of Species by Means of Natural Selection, or the Preservation of Favoured Races in the Struggle for Life*. You might have heard of it. Its critics had many questions, one of which concerned the lack of hard evidence for his proposals. Where, they asked, were the transitional species of which he wrote?

Now here was an ancient animal to give substance to his argument.

About the size of a magpie, *Archaeopteryx* showed enough bird-like traits – wings, feathers, hollow bones – to be given the name *Ürvogel* ('first bird') in German. But it also had distinctive reptilian features: long, bony tail, teeth, clawed fingers. As palaeontologist Hugh Falconer wrote to Darwin in 1863, 'Had the Solnhofen quarries been commissioned – by august command – to turn out a strange being à la Darwin – it could not have executed the behest more handsomely – than in the *Archaeopteryx*.'

Yet while *Archaeopteryx* was clearly Something Important, its exact place in the grand scheme of things remained a conundrum.

Thomas Huxley – a loyal supporter of Darwin's ideas, to the extent that he became known as 'Darwin's bulldog' – made the anatomical connection between *Archaeopteryx* and another recently found fossil from the Solnhofen limestone, a small dinosaur called *Compsognathus* ('elegant jaw'), although he stopped short of suggesting an evolutionary link. For one thing, the two specimens were dated from the same time, so one couldn't be descended from the other. And Huxley's main thrust was the connection between reptiles and birds – dinosaurs were a side issue. Nevertheless, his work has since been held up as foreshadowing future discoveries, partly because of passages like this, from his *Lectures on Evolution* in 1876:

> We have had to stretch the definition of the class of birds so as to include birds with teeth and birds with paw-like fore limbs and long tails. There is no evidence that *Compsognathus* possessed feathers; but, if it did, it would be hard indeed to say whether it should be called a reptilian bird or an avian reptile.

Huxley and others came close to making the connection between birds and dinosaurs, but as time passed, opinion changed, shifting towards the idea that any similarities had to be the result of convergent evolution – the independent evolution of similar traits in unrelated lineages. And gradually the focus shifted away.

In 1926 Gerhard Heilmann's *The Origin of Birds* made the argument that the fused wishbone possessed by birds – the furcula – was absent in dinosaurs. This seemed to deal a death blow to the idea that they were related. Dinosaurs became generally thought of as large, cold-blooded, lumbering beasts that went *rawrrrr*, while *Archaeopteryx* remained unchallenged as 'the first bird' – descended, as Heilmann suggested, from reptiles.

And then in 1964 palaeontologist John Ostrom found a surprising foot in Montana, and everything changed.

The surprising foot had three toes. Two of them were equipped with ordinary claws, but the third claw, sharp and lethal, was held up and away from the ground. This was not a claw for walking on – it was a claw for slashing and tearing. Over the next couple of years Ostrom and his team gathered more than a thousand bones from the site, the remains of at least four individual dinosaurs. Piecing together a picture of the animal that possessed the claw, Ostrom concluded that it was bipedal, small, nimble and fast – a leaping predator that slashed its prey and ate it alive. This was a picture sharply at odds with the prevailing view of dinosaurs as clodhopping thickos. In 1969 he published his description of *Deinonychus antirrhopus* ('counterbalancing terrible claw', the 'counterbalancing' in reference to the long, stiff tail that Ostrom surmised was used like a tightrope walker's balancing pole).

The following year Ostrom made a further discovery. He was visiting the Teylers Museum in Haarlem, the Netherlands, and found himself studying a jumble of bones in two limestone slabs. It was the *Pterodactylus crassipes* fossil mentioned earlier – Hermann von Meyer's 1857 finding. Ostrom found himself unconvinced by something about the bones. They didn't conform to his understanding of pterosaur anatomy. In fact, he realised, they were much more like the bones of the *Deinonychus* he'd just spent five years studying. Holding one of the slabs up to the light, he saw indistinct ridges in the slabs: feathers. This wasn't a pterosaur. It was another *Archaeopteryx*. The first one discovered, in fact. It's just that Hermann von Meyer didn't realise it at the time.

These discoveries triggered what has since been called 'the dinosaur renaissance'. Dinosaurs captured the public's imagination as never before – especially, let's be honest, the violent killy ones.

Over the next decade, Ostrom published a series of papers making explicit the similarities between *Archaeopteryx* and *Deinonychus* – hollow bones, long arms and a host of other anatomical features. Possibly the conclusive argument concerned the feature dismissed by Heilmann in 1926. Ostrom found that some dinosaurs did, after all, have fused wishbones. Take away *Archaeopteryx*'s feathers, and what you had was a dinosaur.

Over the next thirty years, consensus gradually built around Ostrom's controversial hypothesis. And in the 1990s, evidence was found that would put the matter beyond reasonable doubt – and in some abundance.

The first discovery was made in Liaoning, a Chinese province not far from the North Korean border. Fossils abound here, and there is a healthy trade in them – a handy way for impoverished farmers to make some extra money. But the one found by a young farmer, Li Yumin, in August 1996 was of a different order. It was a turkey-sized animal, with a large, bird-like skull and a very long, 64-boned tail, which took its length to about 1.3 metres. Most interestingly, there were also indistinct details along the head, back and tail. The significance of these marks wasn't immediately apparent, and there was some controversy over their identification, but it was soon established with some certainty that they were feathers.

The fossil earned the name *Sinosauropteryx prima* ('first Chinese lizard wing'). Soon there would be further discoveries from the same area, such as *Caudipteryx* ('tail feather') and *Protarchaeopteryx* ('first ancient wing'). These things were flightless, which raised a new question: why did they have feathers, if not to fly? Looking at the different ways modern birds use their feathers threw up several possible explanations, including insulation and display. The notion that many dinosaurs had feathers, developed for various purposes and later co-opted for flight by the ones that evolved into birds, is, despite the opposition of a small and extremely vocal minority of naysayers, now widely accepted. Thomas Huxley had floated the idea 120 years earlier, and now there was conclusive evidence. Birds are dinosaurs.

So my little factoid, smugly trotted out to bemused adults all those years ago, turned out to be only partly true. *Archaeopteryx* did live about 150 million years ago – alongside pterosaurs, which were reaching their peak of diversity at this time, and about 50 million years after the first true dragonflies emerged. And Hermann von Meyer did find the first one – just not when he thought he did. And as to whether it was indeed the first bird . . . well, the jury's still out on that one.

What is a bird?

If you're talking about the ones alive today, this is a relatively straightforward question. The bird – like grief or hope, depending on your preference of author – is the thing with feathers. Even though you'd be forgiven for looking at a kiwi and wondering how exactly it gets away with wearing a hairy poncho and calling it 'plumage', all birds have feathers, and nothing else does. There are other characteristics that go towards defining a bird – toothlessness, beakiness, the lightness and strength of their skeletons – but the first and most straightforward point of identification remains featheriness.

Take it back to the origins, though, and the waters quickly become muddied. There are several stages at which you might look at one of the many feathered dinosaurs of the Mesozoic and say, 'This is where they become bird-like.' What it boils down to is your definition of the word 'bird'. Was *Archaeopteryx,* for all its fame, a bird at all? Feathers and wings are one thing, but it also had teeth and a bony tail, neither of them features of modern birds. And lacking a keeled sternum, it was probably a weak flyer, at best. And what about *Xiaotingia,* another candidate from the Liaoning formation, of similar build and closely related to *Archaeopteryx*? This hen-sized animal lived about 5 million years earlier than its cousin. With flight feathers on all four limbs and tail, it was plausibly a flyer, and when it was found in 2011, serious doubt was cast on *Archaeopteryx*'s status as 'first bird'. Or how about *Anchiornis,* another member of the same group – a shaggy, gangly, crow-sized thing from about the same time? Known from hundreds of specimens, it had feathers all over its body, including its feet. Not as aerodynamic as *Archaeopteryx,* it was most likely a glider or, at best, a weakly flapping flyer.

And then there are things like *Velociraptor* (made famous in the Jurassic Park films) and Ostrom's original find, *Deinonychus,* which sit in another group, the deinonychosaurs. All of them dinosaurs, and all of them showing, to a greater or lesser extent, bird-like characteristics. One of the most intriguing of this group is the pheasant-sized *Microraptor,* from about 125 million years ago. Like *Xiaotingia,* it enjoyed feathers so much it grew them on all four limbs, leading to the widespread use of the description 'four-winged'. While strictly accurate, this description gives the impression of *Microraptor* as some sort of super-flyer – twice the wings, twice the power – but even though all those feathers were long, asymmetrical and extremely similar to the flight feathers of modern birds, it seems to have lacked the chest muscles required for fully powered flight, so those feathered limbs were most likely used for gliding from tree to tree.

Or how about *Jeholornis,* from about the same time, another one with flight feathers, but also a strong shoulder girdle, the only problem with this being that, like *Archaeopteryx,* its alignment meant that the wings couldn't be fully raised for flapping flight.

Then there's my favourite, *Mei long* ('sleeping dragon'), a troodontid from 125 million years ago. This fossil, a juvenile, was killed instantly by volcanic ash and found in a position that will be familiar to anyone who has seen a sleeping duck – with its snout nestling gently under one of its forelimbs. This pose, preserving a dinosaur in eminently birdy behaviour, is just as persuasive as any anatomical evidence for the link between the two.

What we like is to be able to point to one thing – a moment, a species, a defining event – and say, 'There. That's when it all changed.' As the one species we knew about that might possibly fill the role 'first bird', *Archaeopteryx* suited admirably. But as further information has come to light and we have learned more about the path to bird-dom, the situation has become increasingly blurred. Our understanding changes not only with the discovery of new species, but with each re-examination of existing data. By the time you read this, new information or interpretations might well have turned up. The awkward truth is that you could make an argument for or against any number of bird-like animals, based on your own personal definition of the word 'bird', and someone could equally plausibly shoot your argument down in flames based on theirs. The appearance of several potential flyers from different lineages – whether or not those lineages led to the ones we now know as birds – suggests that the statement 'powered flight has evolved four times' is wide of the mark. It seems to have evolved several times in dinosaurs alone.

At one end of the path to birds you have things that are undeniably dinosaurs and equally undeniably not birds. *Tyrannosaurus rex*, for example – probably the most famous member of the theropod dinosaurs from which all birds are descended. We all know that one. Big thing, ruthless killer, comically tiny arms, probably had some feathers. At the other end, the huge variety of birds we have today. Say, for example, the house sparrow – little brown job known over much of the world for its perky chirping. How do you get from one to the other?

Things need to change.

You need to lose some weight, for a start. The lighter you are, the less muscle you need to move yourself through the air, and the less food is required as fuel. The dinosaurs that found themselves on the cusp of turning into birds had already slimmed down quite substantially compared with their Triassic ancestors, and now they took it further.

They dispensed with the bones in the tail, evolving instead a pygostyle – the swollen, fatty, fleshy clump known to poultry eaters as the parson's nose – which holds and controls the tail feathers. They developed big plates on the chest – all the better for anchoring flight muscles – which in turn developed a deep keel to enable more powerful wing strokes. They fused bones in various parts of the body. They replaced teeth with light beaks made of keratin, overcoming the difficulty of processing food by developing efficient gizzards to do the same job. They dispensed with several organs – for example the bladder – and shrank others – birds have very small hearts and livers. In addition, females usually have just one functional ovary (usually the left), and the male testes are also often asymmetrical. The thinness of their bone walls – a characteristic shared with pterosaurs, and indeed their common ancestors – also played a large part in the gradual move towards flight.

Another important aspect of their ability to pursue the high-energy activity of flight is their respiratory system. Mammals have a comparatively inefficient way of breathing – a two-way system that mixes incoming new air with old, resulting in a depletion of oxygen to the bloodstream. We muddle on nonetheless, of course, but we are also quick to fatigue. Birds' lungs are small and comparatively rigid, and are enhanced by a system of, typically, nine air sacs (the actual number varies, depending on the species). They operate a one-way system, the air coming into the lungs through the trachea and bronchi, and then circulating round the air sacs. The advantage of the one-way system is that the rate of oxygen-rich air passing through the system is maintained throughout, which means oxygen is diffused more efficiently into the blood.

Another development, not related to weight saving, but crucial to the history of birds down the line, was what they did with their big toe – or, as it is properly known, the hallux. It might not seem like a big deal, but the reversal of the hallux led to birds being able to perch – and given that over 60 per cent of all birds today, including the humble house sparrow, are perching birds (passerines), it seems to have played a great part in their success.

Last, but by no means least, they perfected feathers. That's probably the biggest one of the lot.

You would be hard pushed to design from scratch something as excellent as a feather. They're light. They're waterproof. They have superb insulating qualities. They form a protective layer against the ravages of sun, wind and rain. They form a shield against spiky foliage. Go fast enough and you can make the air vibrate noisily through them to advertise your brilliance as a potential mate. You can make them all sorts of different colours, to the same purpose. You can, if you're a grebe, ingest your own and feed them to your young. Or, if you're a long-tailed tit, you can use hundreds of them as a nest lining.

Best of all, they can help you fly.

They have their uses for humans, too. We've used feathers as writing implements, body ornaments, tickle sticks, dusters, pillow stuffings, arrow stabilisers, medicines, fish bait, toothpicks, powder puffs, biofuels, art materials, dance accessories and probably quite a few other things that would raise an eyebrow or two. I have been known to use a feather as a sort of worry bead, zipping and unzipping the barbs, smoothing them between finger and thumb in a meditative rhythm until they're back in their original, deeply satisfying form. But maybe that's just me.

The other thing you can do with a feather is just look at it, admire its perfection, and wonder how such a thing came to be.

For a long time it was thought that feathers evolved from scales, the idea being that the scales split and frayed and then split and frayed some more, eventually resulting in the forms we know today. This was orthodoxy, taught everywhere, and accepted as the best explanation. Until, that is, 1999, when Richard Prum, then Associate Professor of Ornithology at the University of Kansas, had a light-bulb moment. Like many brilliant ideas, it now seems obvious and simple. All feathers, he realised, are tubular. Scales are not. So one could not have evolved from the other. Instead, advanced feathers evolved from simple ones. He proposed five stages of feather development, each one more complex than the last.[1] First there is the simple tubular quill, with a single filament and no branches, as worn by some pterosaurs. Stage two sees some branching of the filaments into the kind of downy fluff you find on a chick. The next stage sees the addition of smaller branches and a more organised structure around the central vane. Stage four sees further strengthening of the central vane, and the emergence of interlocking barbules. And finally, stage five: the full asymmetrical modern flight feather, as found on the wing of any flying bird near you – an elegantly curved structure formed of a system of interlocking barbs and barbules around a strong central vane. Each type of feather has its own function, and each was dependent on the existence of the previous one.

From the aerodynamic point of view, the stage-five feathers are of the most interest. Attach a series of them to a bone in an overlapping pattern and you have an efficient aerofoil. The leading edge is thick, the curved surface tapering to a thin trailing edge. The brilliance of the arrangement lies in its flexibility. Held together fairly tightly, the wing forms a smooth upper surface over which the air flows freely, albeit with the inevitable turbulence around the trailing edge and tips. But that solid surface is malleable and adaptable – valuable traits when you want to manoeuvre yourself in the air with the minimum of effort.

With this combination of feather types, birds are equipped not just for flight, but for life. Feathers, in short, are quite the thing. The Wright brothers knew this. They studied birds keenly and worked out that the flexibility afforded by feathers was a key component of birds' success in the air. And while they couldn't imitate the miraculous qualities of feathers, they took that flexibility and tried to apply it to their aircraft designs.

Others were less astute.

An enjoyable half-hour is to be had on YouTube, watching videos of humans' early attempts to fly. Brief clips of indistinct monochrome footage show machines of great ingenuity and variety. The one thing they have in common is that they all – chaotically, heroically, magnificently – fail to fly. There are men jumping off ledges wearing large plywood arms; men in superbly contrived Heath Robinson machines that fall apart the moment they're switched on; men (it's always men) sitting in contraptions that might, if developed properly and wisely with extended periods of investment, conceivably turn into moderately enjoyable fairground rides.

The commonest failing seems to be related to a basic misunderstanding of the nature of flapping. Up and down they go, those wings. Up and down and up and down and up and down, and yet the putative flyer remains resolutely earthbound. Part of the problem is of course that they're simply not producing anything like the amount of lift required to get them off the ground. But even if they do somehow manage that, any lift potentially produced by the downstroke would be immediately and completely negated by the upstroke. All they're doing is bouncing up and down on the spot. That straightforward 'up–down' movement might seem intuitive, but birds' wing strokes are more sophisticated and variable than that, as we shall see.

This wasn't the only problem with these designs, of course. Most of them, to our sophisticated modern eyes, were quite clearly doomed to failure. Yet they represent honest and earnest attempts to solve a problem that must at the time have seemed tantalisingly within reach.

This period, when the technology for flight was in its infancy, is not dissimilar to the beginning of the development of flight in birds. Because for all the *Archaeopteryx* and *Microraptor* and *Xiaotingia* and other creatures that were in possession of at least a half-workable strategy for staying in the air, there were many others that might superficially have looked the part – wings, feathers – but when it came to it, they couldn't quite do it. Wings, after all, developed before flight, and had other purposes before they became useful propellers. A pair of wings can be used for stabilisation while running, or they can help you get up a tree, especially if they have claws for gripping. Spreading wings suddenly can scare off predators or be used in display to attract a mate. Or you might want to fan them in an umbrella shape to shield your food from opportunist scavengers (chip-eaters of the British seaside, take note).

Whatever their uses, they became more than useful for birds when eventually co-opted into what we now think of as their primary purpose. How that all started remains a matter of great debate, each hypothesis having its proponents and critics.

The cursorial or 'ground-up' model has it that proto-birds started on the ground, running fast. They developed their feathered arms to help with stability, which enabled them to go faster. Then a jump or two – perhaps a bit of a glide to catch that mosquito, and why not flap a bit to make the running easier? The next stage might have been a sort of flappy glide, leading, in time, to fully powered flight.

Counter to this is the arboreal or 'trees-down' model. A tree-dwelling feathered dinosaur would first jump out of its tree. Then, with the help of the aerodynamic surfaces on its arms, some gentle gliding to help it go further. A bit of a flap to prolong the journey, and before you know it, the arms have become proper aerofoils and the world is your oyster.

This debate has many similarities with the debate about how insects first got going – opposing arguments, each denying the merits of the other. Each argument has its difficulties.

In some ways the arboreal model is a more intuitive explanation because it relies on mild manipulation of gravity rather than outright defiance of it. But the transition from gliding to flapping is tricky to explain because at some point it requires the development of flight muscles, which add weight and make flight more difficult, and there's no good reason for evolution to favour that kind of retrograde step. Meanwhile, the cursorial model depends on birds having enough power to launch themselves. Looking at *Archaeopteryx*, this seems unlikely – its legs wouldn't have had that kind of strength.

As usual, my inclination, when these hypotheses are explained to me, is to nod along in agreement until the next one is brought out, at which point I think, 'Oh yes, that does make more sense, actually.' And then someone comes up with an alternative hypothesis and I think, 'Oh.'

Enter Ken Dial, with an explanation that encourages the response, 'OK, now you're talking': the so-called wing-assisted incline running model (WAIR for short).[2] Dial, professor of Biology at the University of Montana, noticed that when flightless partridge chicks wanted to escape danger, they scrambled up trees, using their poorly developed wings to help. The chicks were not yet capable of flight, but were putting their wings to good use nevertheless.

This concept will be familiar to anyone who has encountered partridges on the winding country lanes of Great Britain. A red-legged partridge, pursued by a car, will run first. It seems as if they will go to any lengths not to fly. At times it feels like a death wish. They are, to be fair, pretty nifty runners, even if, in their panic, they often fail to see the benefits of just running to the side instead of straight on. They are, of course, capable of flight, although they mostly use it only in short bursts, such as when being pursued by a Škoda Octavia in the country lanes of Suffolk.

Dial constructed an experiment. He encouraged partridge chicks towards a ramp and observed their behaviour as he adjusted the angle. He found that up to about 45 degrees, the chicks made their way up the ramp without using their wings to help. Above that, they started flapping, and as the angle became tighter the depth of their wing strokes also increased. Dial's own analogy likens the wings to the spoiler on a racing car. The idea is that early birds had poorly developed wings like those of partridge chicks, so they used them in exactly that way until they evolved into something more useful for the beginnings of sustained flight.

The elegance of this suggestion is that it works as a kind of halfway house between the arboreal and cursorial. It provides an explanation for the development of flapping flight, with a neat reason for the existence of half-wings. And whether you subscribe to the arboreal or cursorial hypotheses, it is consistent with incremental changes from not flying to flying.

However it arose, once everything came together the evolution of birds accelerated, and by the time of the late Cretaceous, about 66 million years ago, the skies were buzzing. Insects, lest we forget, hadn't stopped radiating and diversifying; pterosaurs had attained almost unimaginable sizes; and feathered flyers of all descriptions were slotting into the niches unoccupied by the pterosaurs.

The dinosaur species that developed feathers and wings and took to the air had increasingly come to resemble things we would recognise as birds. Regardless of where you place them in the family tree, and whatever you call them, you would find a healthy variety of flying things, a lot of them looking like birds and behaving in suspiciously bird-like ways. Life in the air was thriving as never before or since.

And then it all went horribly wrong.

9

THE PENGUIN

A beach. Clear blue water. Warm sun, doing its thing. You could be in paradise.

But you're not. You're in London NW1. The beach is artificial, the water (450,000 litres of it) carefully regulated to ensure the continued health of its inhabitants, and the sun is about to disappear behind a looming cloud.

On the plus side, there's an underwater viewing area. And on the double plus side, there are penguins. Humboldt penguins, to be precise. *Spheniscus humboldti.* Loads of them. Endemic to the west coast of South America, says the information board. Nests in colonies on islands and rocky coastal stretches. Eats fish and squid. Status: vulnerable. Flightless.

A flightless bird is a conundrum. Why, having gone to all the time, trouble and effort of evolving flight, would you then abandon it? Surely flight is the pinnacle, the *ne plus ultra* of animal locomotion?

Try telling a Humboldt penguin. It's one of sixty or so living species to have abandoned the air in favour of a terrestrial (and aquatic) lifestyle. It doesn't know what it's missing but has compensated with other skills.

There's one within about two metres of me. Black and white, with a patch of pink around the face. Knee high to me. Head high to a three-year-old. If it weren't for the perspex screen I could almost reach out and touch it. It gives me a look, beak raised, as if to say, 'Don't even think about it, pal.' Then it slides off its rock and eases into the pool, moving smoothly and powerfully through the water with economical strokes of its flippers.

Standing nearby, a three-year-old – head high to a Humboldt penguin – is giddy with glee.

'Pen-ging! Pen-ging!'

Close enough.

When an asteroid hits your planet, there are many ways to die.

If you're anywhere near the initial impact – say within a few hundred kilometres – you will be instantly vaporised. This – instant, painless death – might be the best option. If you're a bit further away, you might be suffocated by the phenomenal outpouring of ash generated by the impact. Live near any coastline and you might be engulfed by the resulting tsunamis. Somehow negotiate these hazards, and you'll be faced with another, in the form of molten rock raining back down on the earth. If it doesn't hit you, it heats the ground to unbearable temperatures.

Perhaps you survive the first death wave. Perhaps you manage to burrow into a safe place or are far away on the other side of the planet. Respite is temporary. The impact has generated so much debris – choking fine ash and dust – that it spreads across the world, blotting out the sun for years on end. Without light, plants die; without plants, there is no food; without food, there is death. Acid rain only adds to your woes.

If you live in the sea, things are only marginally better. The rise in temperature reduces oxygen levels. This is not good. It amounts to a slow asphyxiation. All those tiny things at the bottom of the food chain are done for, with inevitable knock-on consequences.

Life is fragile.

The asteroid that hit Chicxulub in the Yucatán peninsula in Mexico just over 66 million years ago was about 10 km across. The effect it had was roughly as described above: destruction on an unimaginable scale. While it's now generally accepted that the asteroid was the main instigator of the K–Pg extinction event, there is another possible contributor to the carnage. Around this time, there had been sporadic volcanic activity in a large area of India – about half a million square kilometres – now known as the Deccan Traps. There is a hypothesis that the asteroid impact triggered an increase in the already significant sulphurous emissions brought on by this volcanic activity, creating a perfect storm of global destruction.

Whatever the reasons, the effect was devastating. Estimates vary, but around 70–75 per cent of all plant and animal species are thought to have gone extinct. And those species that did maintain a population in the aftermath were badly hit too. The upside for the survivors was the opportunity afforded by the new status of the world. Ecological niches abounded, just waiting for the right species to fill them.

So, what were the right species? How do you survive a mass extinction?

Firstly, don't be big. Anything over about 25 kg, with long life cycles and slow reproduction rates, was toast. If you were a specialist, reliant on one particular food source or habitat for your survival, that reduced your chances. The more complex the ecosystem, the slenderer the thread by which it clings to life. Coral reefs and tropical rainforests do not do well in mass extinctions.

Among many other things, pterosaurs were done for. It's not clear whether they were already in decline and the asteroid and its aftermath merely delivered the *coup de grâce*, or whether they were thriving and cut off in their prime. What does seem to be the case is that they were not well served by their own evolutionary development. For all that they had thrived for millions of years, a combination of their lack of diversity and their size left them vulnerable to extinction. The pterosaur families that were around at the time – pteranodontids, nyctosaurids and azhdarchids – were generally large, certainly in comparison with birds. *Quetzalcoatlus, Hatzegopteryx, Pteranodon* – awe-inspiring animals that pushed the limits of what was possible in the air. But they and their cohorts had big flappy wings that didn't fold neatly away so they could find somewhere to shelter. The nature of their eggs won't have helped. They were soft-shelled and not incubated, so hatching rates will have fallen with the drop in temperature in the immediate aftermath of the asteroid impact.

Birds, on the other hand – smaller, with fully foldable wings and hard-shelled eggs – were generalists, able to survive on a varied diet of whatever they could find. They not only made it through, but ultimately thrived, while the other dinosaurs famously perished. These extinctions didn't happen overnight – the whole thing played out over thousands of years – but once they were gone, they were gone.

The birds didn't survive unscathed, however. Also abundant at the time was a group called the enantiornitheans – 'opposite birds', the 'opposite' referring to the reversed orientation of their shoulder bones, a distinction that either made them weaker flyers than birds or just as strong, depending on who you listen to. They were a diverse group, resembling modern birds in most external features. For reasons that remain unclear, they bit the dust. So, too, did the hesperornitheans ('Western birds'), a group of mostly flightless, strong-diving, cormorant-like birds that were the only ones to colonise the oceans in this period.

The ones that were left belonged to two clades: paleognaths ('old jaw') and neognaths ('new jaw'). Their lineages diverged about 100 million years ago, and the neognaths went on to split, between 10 and 15 million years later, into two more clades, the Galloanserae ('rooster-geese', containing the ancestors of today's chickens, ducks and all their relatives) and the Neoaves ('new birds', containing the ancestors of literally everything else).

The point arising from this is that the ancestors of modern birds started diversifying well before the extinction, living alongside their non-avian dinosaur cousins. But while their diversification in that time was successful enough, it was nothing compared to what happened once the dust had settled on the extinction. In the 15 million years that followed, the surviving birds underwent an extraordinary pulse of evolution, with the result that by 50 million years ago, nearly all the orders from which today's birds are derived had established themselves. And from there they radiated and grew to colonise the whole world.

Birds are everywhere. On all the continents, in all the habitats. They live in the hottest, coldest, wettest, driest, highest and lowest places on Earth. Some live alongside us, making their homes in ours. Others opt wisely to stay as far away from us as avianly possible. From ocean to forest, desert to ice floe, garden to mountain – you are rarely far from a bird.

There are tiny ones, huge ones, chunky ones, slender ones, long ones, short ones. There are birds of colouring so extravagant you need sunglasses to look at them, and others of a quite remarkable drabness. They are readily observable with the naked eye – even the smallest, the bee hummingbird, is 5 cm long. They also provide a sound world of extraordinary profusion and variety. No wonder so many humans are drawn to them.

According to the eBird/Clements checklist* there are 41 orders of bird, divided into 248 families. A total, at the last count, of 10,824 species. This number has gone up in recent years, not necessarily because large numbers of previously unknown species are being found – these discoveries might amount to just a few each year – but because birds previously thought to be one species turn out, on examination of DNA, to be two or more. In the early 1980s the number was reckoned to be between 8,000 and 9,000, and a 2016 study estimated that the number might swell to as many as 18,000 with new taxonomic research being done

* There are four generally accepted bird checklists, each with its own approach to taxonomy, and each affording species status to different sets of birds. But even though the world of taxonomy is fraught with complexities, the four lists can at least agree on 87 per cent of species.

all the time. Whether the number will get as high as that rather depends on whether you are a 'lumper' or a 'splitter'. To illustrate these terms with an example from everyday life, a 'lumper' would regard all Magnum ice-creams as the species 'Magnum', regardless of flavour; a splitter would deem the Double Raspberry Magnum and the Salted Caramel & Glazed Almond Magnum two separate species.

Glad to have cleared that up for you.

The sixty or so flightless species out of those 10,824 come in a range of sizes, from endearingly small (the 15-cm Inaccessible Island rail) to terrifyingly huge (the 275-cm common ostrich). While they might not be especially pleased to hear it, we often find them both comical and endearing. Perhaps for this reason, flightless birds are among the best known of all. You might never have heard of cotingas, buttonquail or motmots, might be completely unfamiliar with the appearance of a caracara, antbird or potoo, but I have a fiver here that says you know what a penguin looks like.

Penguins are not like other birds. Our Humboldt penguin at London Zoo was a representative sample of the smaller ones, but perhaps the most famous of the eighteen commonly recognised species in the family Spheniscidae – thanks at least in part to the film *March of the Penguins* – is the Emperor penguin. Strange to think that these large, waddling, entirely adorable hulks – one metre tall and up to 45 kg in weight – are, however distantly, related to the flitty blue tits and robins denuding your garden feeder of sunflower seeds. Their dignified, upright posture is quite unbirdlike, their black-and-white plumage leading to comparisons with waiters or footmen. Their wings are clearly unsuitable for flying – stiff, slender appendages, disproportionately small for their body size. And there's a sleekness to their plumage that in some ways resembles a mammal's pelt or fur.

There's more to that plumage than at first meets the eye. The smooth outer layer – black on the back to absorb heat, white on the front for camouflage – is made up of short, overlapping contour feathers that serve them well for both waterproofing and insulation. Underneath, their defences against the extreme cold of the Antarctic continue. A layer of fat under the skin is normal procedure for any animal wanting to withstand temperatures as low as –40°C, and Emperor penguins duly have a good wodge of it contributing to their blubbery appearance. But between the skin and the contour feathers is the most important bit – a dense layer of dual-purpose downy feathers called plumules. Plumules are excellent insulators – it's that delicious fluffiness – and they are backed up by another kind of feather known as afterfeathers. These too are small, and they are attached to the base of the main feathers rather than directly to the skin, like the plumules.[1] This complex network of fluff is such an efficient insulating layer that manufacturers of outdoor gear are keen to unpick its secrets.

The plumules' second function becomes clear once the birds are in the water. This is where penguins spend about half their time, and the transformation from clumsy stumbler to sleek diving machine is startling. Those apparently feeble flippers are extremely well adapted for propelling the piebald blubber tubes through water, and they use them to good effect, sometimes descending to depths of 500 metres and staying underwater for up to twenty minutes. Unlike other birds, penguins have solid bones, the better to help them withstand the intense water pressure at that depth. But it's when they come back up that the genius of the plumules is revealed. It's one thing being able to dive to great depths and generally manoeuvre yourself nimbly and efficiently in the water, and penguins are adept at that. But it's quite another to generate enough upwards momentum to launch yourself clear of the water and return to solid land. What you need is to reduce drag, and this is where the mesh of plumules comes in, trapping tiny bubbles of air to form a buoyant coat surrounding the penguin. Spending time in the deep reduces the volume of that air coat by up to three-quarters, so it needs replenishing. First, the penguins swim slowly up to the surface and give themselves a bit of a preen. Then they descend again before making the final run to launch themselves out of the water. The preening reinstates the bubbles, which are released as the penguin plunges upwards towards the surface, leaving a wake of bubbles behind it like a vapour trail. The drag reduction enables them to swim at nearly twice the usual speed as they approach the surface. Out they pop, like bullets from a nerf gun, landing in slightly undignified fashion on the ice – or, sometimes, on the back of another penguin.

No wonder we love them.

The hypothesis for how penguins evolved to be flightless is much as you'd expect. Go back to shortly before the K–Pg extinction, and there is a large clade of seabirds, including the ancestors of albatrosses, divers, shearwaters and penguins. The penguin lineage split from this group, survived the extinction and took it from there. It's thought the massive loss of marine life in the extinction left a vacuum, which the penguin ancestors exploited, adapting their bodies extraordinarily quickly to a flightless aquatic lifestyle. The earliest fossil we have, *Waimanu manneringi* ('Mannering's water bird' – *wai* in Māori meaning 'water', and *manu* meaning 'bird'), dates from about 62 million years ago on New Zealand's South Island. Not yet fully penguiny – while its bones were of a density consistent with flightlessness, its wings still resembled wings rather than flippers – it was nevertheless recognisable as something that might one day, if it really put in the hours, become a penguin.

Many, many hours passed between *Waimanu* and today's penguins, and in that time they have become universally flightless. The break-up of the Gondwanan landmass created new conditions in the southern seas. New islands appeared, with extensive coastlines where the birds could form colonies safe from terrestrial predators, and new oceanic currents developed to help dispersal across the circumpolar region. Mostly isolated from terrestrial predators, and well placed to take advantage of the many fishing resources that arose as the polar ice cap formed and the surrounding waters cooled and became more nutrient-rich, they gradually evolved away from flying and towards diving. The wings that once propelled their ancestors through the air adapted to the different challenge of water displacement, becoming smaller, stronger and stiffer – great for swimming, but less good for flying. At some point the balance tipped – the energy costs of flight became too big, and the benefits of a flightless lifestyle outweighed them, so their wings converted to flippers once and for all.

Penguins have remained almost entirely confined to the southern hemisphere – the only exception is the Galápagos penguin. We might associate them with the ice of Antarctica, the home of the better known (and larger) species, but plenty of them are temperate creatures (especially the smaller ones), with representatives on all the major southern landmasses. Despite the challenges brought by human intervention, and not least climate change, flightlessness and a semi-aquatic lifestyle have served them extremely well.

Today, penguins make up about a third of the flightless species on the planet. Accounting for another quarter is the group descending from the paleognaths that sit at the base of all modern bird lineages. They're the ratites – the oldest (and, some might say, oddest) birds on the planet. The name derives from the Latin *ratis,* meaning 'raft', and refers to the flat shape of their breastbone. They count among their number ostriches, emus, rheas, kiwis and cassowaries, each group living in different parts of the world – ostriches in Africa, emus in Australia, kiwis in New Zealand, cassowaries in New Guinea and north Australia, and rheas in South America.

It was Darwin – who else? – who proposed that, despite their wide geographical dispersal, these birds all belonged in the same family. They had plenty in common – underdeveloped breast muscles, a flat sternum with no keel for anchoring flight muscles, strong legs (terrifyingly so in the case of ostriches), simple wings and unvaned feathers.

The question remained: how did they reach their current homes?

The standard explanation was that they all evolved from a common flightless ancestor, resident of Gondwana. When that huge landmass broke up between 130 and 50 million years ago, they were each isolated on separate bits and carried off to their current homes. This explanation seemed intuitive and logical – how else would flightless birds cross the seas?

Enter the tinamous.

Tinamous are endearing fowl-like creatures found in Central and South America. Compact and round-bodied, with cryptic plumage and slim necks, they are shy ground-dwellers that – like the gentleman of legend who knows how to play the trombone but doesn't – have the ability to fly but don't often choose to use it. For a long time the tinamous were thought to be closely related to ratites, but not members of the club, standing outside with their noses pressed forlornly against the window. This assessment was based partly on their morphology – they have a keeled sternum and preen glands, both features lacking in ratites – but also on their ability to fly. According to the 'flightless ancestor' hypothesis, this alone would place them outside the ratite family. On the other hand, as our old friend Thomas Huxley noticed, they share one feature with the ratites – reptilian palates. Odd.

By these assessments they would surely belong with equal validity both within the ratites and outside. Schrödinger's tinamous.

Then DNA analysis came into the equation. A 2014 study delved deeply into the relationships between the various birds and came up with some surprising results. It showed that the closest relatives of the South American tinamous were the extinct moas of New Zealand, which were definitely ratites. This places tinamous firmly in the ratite club, drawing up a chair in the main dining room and perusing the wine list. This suggests that the common ancestor of the ratites was not flightless. Rather, they dispersed by flying to their various destinations, then flightlessness evolved separately multiple times via convergent evolution, with tinamous the sole group retaining the ability to fly. The alternative explanation, that tinamous evolved flight from an ancestor that had become flightless, is, while not impossible, considered overwhelmingly unlikely.

It's customary among bird people to refer to certain birds as 'the most prehistoric of them all', based on their appearance and our notion of what things looked like back in the day. Common contenders for this title include the grey heron, visible throughout Britain at the waterside, patiently stalking its prey, its scraggly beard and dagger bill a dramatic sight. Or cormorants, those beaky waterbirds with the waterloggable feathers. Head for Africa and the extraordinary appearance of the shoebill – one of the more accurately named birds in the world – conjures images of a world before we were around.

While the visual aspect of those birds offers a reminder of their ancestry, the honour of 'most primitive birds' in fact goes to the ratites – their lineage being traceable directly back to well before the K–Pg extinction. Look at an emu's foot for confirmation. Three heavy, scaly toes that wouldn't look out of place in Jurassic Park. If it weren't for their feathers and the vestigial wings, you might look at these creatures and wonder where they fit in the tree of life. They are in many ways unbird-like.

How exactly they all became flightless remains one of those hotly disputed topics we've visited so often in this book. The main hypothesis is similar to the one applied to the penguins – lack of predator threat. Why fly when there's nothing to fly away from? In the aftermath of the extinction, with small creatures abounding and an absence of larger predators, an ecological niche for big birds sprang up. Make yourself big, learn to feed on grassy vegetation – relatively abundant compared with the forests that fared so badly – and you can thrive. That lack of predators means you lose first the necessity and then the ability to fly.

The ratites have a colourful history. Among their number, now extinct, were the spectacularly large elephant birds of Madagascar. Measuring up to three metres tall, and weighing as much as 780 kg, the largest of the family, *Vorombe titan,* is reckoned to be the largest bird ever to have lived. Its eggs were no less impressive – each one yielding approximately fifty omelettes.* Little wonder that they proved so popular with Madagascar's human settlers.

All was comparatively well for the elephant birds, living on Madagascar for millions of years without a care in the world, until humans turned up and things took a turn for the worse. It's thought that hunting played its part, along with habitat loss, in the extinction of the last remaining elephant bird species, *Aepyornis hildebrandti,* between 1,300 and 1,500 years ago. A similar fate befell the moas of New Zealand, hunted to extinction by Polynesian settlers.

We have a worrying habit of doing things like that. We killed the elephant birds. We killed the moas. Famously, we killed the dodo. But we haven't – so far, at least – killed the largest of the extant ratites, the ostrich.

The variety of life means that to say something 'looks odd' requires some refinement. Your 'odd' might be something else's 'normal'. But even taking into consideration all those caveats, and without in any way judging or criticising, we can state, definitively and as fact, one thing.

Ostriches are weird.

It's weird how weird they are. Huge for a bird, long-necked and long-legged, with thick, powerful thighs peeping out from under a veil of black feathers. The thighs of a runner. And that, of course, is what ostriches are, capable of sustained travel at 50 km/h, gusting to 70.

* If, that is, you make an omelette with three hen's eggs, like a sensible person.

Ostriches are also a good example of a counterintuitive aspect of bird anatomy. Those legs articulate at what you assume is the knee – simply because it comes in the same sort of place as the human knee. But we're mammalian bipeds whose norm is to walk plantigrade, with our soles flat on the ground (a feature we share, incidentally, with penguins, which might be a factor in our fondness for them). Most birds are digitigrade – they walk on their toes. And the articulation faces the opposite direction to ours. So that knee is not a knee; it's an ankle.

It's also of some surprise to discover that, unlike the other members of their clan, they have fairly substantial wings, spanning about two metres. My, what large wings you have. All the better for . . . well, not flying, as a matter of fact.

Why would you need such things if you're not going to use them for flying? Penguins repurposed theirs for swimming. Ostriches have no such need for them. Surely, having fallen into disuse, they would wither over the generations and finally disappear? Well, it turns out that these wings do have a vestigial purpose. They're effective stabilisers and steering aids, enabling more manoeuvrability than you'd have any right to expect from a fast, heavy bird. They can also be used as a sort of parachute, to help with braking. Away from the running arena, they are used in courtship displays, and can play a role in thermoregulation, the birds using them as cover for their feather-free legs as and when necessary.

Kiwis did something different. They're outliers in the ratite family – staying relatively small rather than growing – and lost their wings almost completely, the only remaining evidence of ancestral flight reduced to tiny vestigial nubs. They also buck the family trend for diurnal herbivory by being nocturnal insectivores. All five kiwi species are endemic to New Zealand, and, like their long-deceased elephant bird cousins, have largely been victims of human activity, all five currently residing in the 'vulnerable' or 'near threatened' conservation categories. Part of the problem has been deforestation, although kiwis are now protected in large reserves, but it's their flightlessness that has proved instrumental in their struggles. And, yes, it comes down to humans again. The flightless birds lived in their habitat for aeons, safe in the knowledge that they wouldn't be eaten by hungry mammals. We arrived, bringing cats and dogs, which hunt and kill creatures that simply don't have an effective escape strategy.

Permanent flightlessness is one thing, but some birds adopt it temporarily as part of their annual cycle. All birds moult. Feathers are amazingly resilient, but they do wear out with use, and are also vulnerable to feather lice. Staggered moulting is the norm for most birds, the feathers replenishing over a few weeks, so that the bird can retain almost full plumage throughout. The flight feathers are particularly important, and these are replaced one by one to ensure the bird retains the power of flight. In large birds of prey the process is visible from underneath as they fly overhead, the symmetrical gaps in the wings as visible as missing teeth.[2]

Ducks, for reasons best known to themselves, do it differently. They opt for the quick, clean approach, replacing all their primaries at once. This is quicker, but during the moult they're rendered flightless. In some species, the males go into 'eclipse', their gaudy breeding plumage replaced by drabber feathers so they don't stick out from the crowd. Wetlands are generally safe places to undergo this metamorphosis, but flightlessness does have obvious dangers. To make themselves less vulnerable to attack, a dabbling feeder will tend to make sure it's near plentiful vegetative cover in which they can hide from predators; diving ducks do the opposite, sticking to large expanses of open water. And some species make short migrations beforehand, to areas with abundant food to fuel the energy-intensive activity of moulting.

Flightlessness is a fascinating aberration, a niche occupied by a small but select group of outliers, most of which have adopted it successfully for millions of years. All that effort to get yourself in the air in the first place, only for subsequent generations to decide they don't want or need it after all. Developing the capacity for flight requires a strong incentive, but so does abandoning it. And once it's gone, it's gone. As far as we know, nothing has made the progression from flight to flightlessness and then back to flight again.

Flightless birds hold a unique charm. We identify with them, anthropomorphise them readily – more, perhaps, than other birds. Nowhere is this chiming with the human spirit more evident than in a clip from Werner Herzog's documentary *Encounters at the End of the World*. An Adélie penguin has separated from its colony as they make for their feeding grounds. It stands alone, then, apparently making a decision, turns inland and waddles gamely – upright, like a human – across the vast expanse of ice, towards distant mountains. The bleakness of both the landscape and the penguin's future is emphasised by a soundtrack of resonant, portentous choral music.* Over it, Herzog's inimitable commentary evokes the penguin's loneliness and vulnerability as it heads towards certain death. He asks one short, potent question, familiar to anyone who has ever contemplated the mysteries of life.

* Bortnyansky's *Retche Gospod Gospodevi Moyemu*.

'But why?'

We might thrill to the spectacle of birds of all shapes and sizes in flight, be fascinated by their behaviour, wonder at their mastery of a skill that will always elude us – but when you want to evoke the eternal mysteries of human existence, you can't beat the hopelessness of Herzog's nihilist penguin.

10

THE GOOSE

There has to be a good reason for me to get up at 4.30 a.m. Geese are a good reason. Plus, I get to have Second Breakfast.

Some people live where there are geese in winter. I don't, so every autumn I visit at least one place where they are. The geese in question aren't the Canada or Egyptian geese readily available in my local park. Those are fine enough in their own way but lack the wild edge. These are pink-footed geese – 'pinks' to their many fans. Visitors from Greenland and Iceland, their arrival in October is a sign that proper autumn has begun. The UK population is estimated to be more than a quarter of a million, and if you find the right place it can feel as if they're all there at once.

The RSPB reserve at Snettisham in Norfolk is the right place. They roost overnight on the broad expanse of mudflats on the Wash, leaving at sunrise to head inland, where they will find a field and eat the bejesus out of piles of sugar beet. Sunrise is 7 a.m., first light about half an hour before that. It's a drive to the reserve and then a walk to the watch point, and I'd rather stand in the freezing dark for forty-five minutes than arrive just after the geese have buggered off, so am leaving nothing to chance.

It's dark on the trail. Faint shadows of hedges. Tick of robin. Shout of wren. Strangled squawk of gull. I reach the watch point and wait. As gloom morphs into daylight, features gradually make themselves visible, and soon the faint flickerings of activity in the dark reveal themselves as birds. Lots of them. Waders – knot, dunlin, godwits, oystercatchers – furtling around in the mud, probing for invertebrates. Today is not a day for the spectacular whirling flights that draw admiring crowds – you need a high tide for that, when lack of space on the mud forces thousands of birds into the air – but their quiet industry has a pleasing rhythm to it nevertheless. I'm also quietly enjoying the solitude. The feeling of space is palpable. Huge sky, expanse of mud. A sense that you are at the end of something, looking out and beyond into eternity.

Far in the distance, the top layer of the mudflats shifts, rising in a sheet as if slowly peeled back by an invisible hand. Only after half a minute or so, as it billows gently and pixels begin to appear around the edges, does it become clear that this single entity is thousands of geese. And now it's splintering into discernible groups and I hear their conversation, a babble of high-pitched honks – raucous but not jarring – as they move towards me. 'Here! Here! Here!'

They gain height, separating into a series of jagged Vs silhouetted against the morning sky. Over my head they come – wave after straggly wave of them, calling and straining, the sight and sound swelling then ebbing until the last of them disappears from view over the horizon.

The pinks are here. Autumn can begin. Second Breakfast time.

Rewind. Just a few years. It is the late Cretaceous, approximately 66.8 million years ago. The asteroid will hit in 700,000 or so years' time. Dinosaurs are doing just fine, thanks, and huge pterosaurs share the airspace with a variety of bird-like creatures.

One of the latter is dabbling around in the warm and shallow waters of the sea somewhere near what is now the Belgian border with the Netherlands. If you saw it today you'd think it vaguely familiar, yet not. A smallish bird. A bit chickeny in some ways, especially around the head. But distinctly goosey legs, perhaps with a hint of pheasant in the body.

This is *Asteriornis** *maastrichtensis*, dubbed 'Wonderchicken' by the scientists who described it in the early 2000s. It is the earliest known member of the Neornithes – 'modern' birds – and is therefore significant. Unlike a lot of the other feathered creatures knocking about the place at the time – those toothed enantiornitheans, for example – *Asteriornis* was closely related to birds alive today. Not quite their last common ancestor, but probably high on its Christmas card list.

Of equal interest is a fossil from around the same time and the other side of the world. *Vegavis*† *iaai* seems, from its delicate remains, to have been more like a diver ('loon' if you're in North

* Named after Asteria, one of the Titans, who according to legend transformed herself into a quail.

† 'Bird from Vega', Vega being the island on the Antarctic Peninsula where the fossil was found in 1992.

America) than a goose, with strong feet set at the back of its body, the better for propelling it under water. Tantalisingly, it seems, from the structure of its syrinx (the avian vocal organ, equivalent to our larynx), to have been capable of goose-like honking sounds.

Asteriornis and *Vegavis* are representatives from a narrow window in history: they're modern birds from the time of the non-avian dinosaurs – and their existence shows that the diversification of birds was already under way before the asteroid hit. Not surprisingly, they are therefore of considerable interest to those concerned with nailing down exactly what happened when. Their precise position is still a matter of considerable debate, but 'in or near the Galloanserae' (the clade informally and broadly known as fowl) is a relatively uncontroversial summary.

Time passed, as it does. Birds diversified, as they do. Out of the Galloanserae came two orders: the Galliformes (chickens, turkeys, pheasants and so on) and the Anseriformes (ducks, geese, swans, screamers and – now represented by just a single species – the confusingly named magpie geese). Of the two, it was the Anseriformes that developed the greater skill in the flying department – although there are, inevitably, exceptions. Two species of teal, confined to their island homes to the south of New Zealand, became flightless, as sometimes happens with island species. Meanwhile four species of South American steamer duck use their shrunken wings in tandem with frantic pedalling to run across the water when alarmed – the resemblance to a paddle steamer giving them their name. As compensation for their flightlessness, they are notably aggressive in defence against potential predators.

These exceptions aside, Anseriformes are often strong flyers, and the family's general body plan, from the 26-cm cotton teal to the 1.8-metre trumpeter swan, is stocky and powerful, with fast-beating pointed wings. Many of them put these physical attributes to especially good use, undertaking lengthy journeys twice a year in a manifestation of one of the great wonders of the natural world: migration.

If migration in the insect world is a minority activity – albeit more common than many imagine – in birds it is commonplace, with about half the world's species practising it in some form. The variations of distance, direction and motivation within the term 'migration' are significant, but as an all-encompassing word it is taken to mean an annual return journey – one way in spring, the other in autumn. And while the life cycle of insects means that such journeys – our painted lady springs to mind – are multigenerational, birds do it all by themselves. In some cases, as we'll see, this means journeys of quite staggering endurance and stamina. But even the comparatively short journeys undertaken by millions of birds each year contain an element of risk. That risk, naturally, is outweighed by the overall benefit – otherwise the whole process would be disadvantageous – and that benefit, and therefore the primary driver of migration, is most often food. Other resources play their part – the availability of mates or nesting sites – but mostly it's about food.

Take, for example, our pink-footed goose. This is one of many species that spend the summer breeding season in the far north. In their case the range is fairly narrow, the bulk of the population breeding in Iceland, Greenland and Svalbard. Conditions there suit them perfectly. They're ground nesters, and the relative absence of predators in the far north means their eggs and chicks enjoy as much safety as any wild creature can expect. There is constant daylight, so they can forage all day. The abundance of vegetation – they like sedges, mosses and lichens – means they're in good condition to produce numerous healthy chicks.

Then winter comes. An Arctic winter is harsh indeed. There is less food, and fewer daylight hours in which to find it. Time to move. The Svalbard population heads almost due south, mostly ending up on the west coasts of Denmark, Germany, the Netherlands and Belgium. The Iceland/Greenland population goes south-east, the bulk of them wintering anywhere in the UK from Norfolk northwards. There is very little crossover between the two populations, the flyways forming a V shape on the map.

There are multiple variations on this pattern – far north in the summer, south in winter. A favourite example is the sanderling – a plump little bird whose scuttling shoreline foraging will be familiar to anyone who has ever seen the Pixar short *Piper.* In the summer they range from Canada to Siberia, and when the time comes they disperse far and wide, hanging around beaches and mudflats, bringing a little burst of joy to anyone who happens on their idiosyncratic wind-up-toy feeding behaviour. You might see them in any coastal region: California, Argentina, Portugal, India, Australia, you name it. They are a true long-distance migrant.

The north–south pattern also applies to birds that spend their summers in Britain. Breed here, move south for the non-breeding season – for some species, such as swallows, that might be as far as South Africa. And while there is often an element of east–west drift in these movements – especially when birds breed in the middle of a landmass and move to the warmer edges for the non-breeding season – the prevailing direction is overwhelmingly north–south.

What this amounts to is a biannual shift in the distribution of birds across the planet. And with times of departure overlapping through the year, the result is an almost constant movement – albeit with peaks in spring and autumn, and troughs in midsummer and midwinter. Whatever the time of year, something, somewhere, is on the move. And nearly all of it takes place out of sight.

These journeys – especially the long ones – capture the imagination, and there's a temptation to romanticise them further with images of lone birds doggedly flying through all conditions over vast expanses of water or desert – the loneliness of the long-distance migrant. But birds often undertake these journeys in groups. There are obvious advantages to this strategy. Through vocal communications the group might find it easier to find food sources, warn each other about predators, and stay on the right track. But even better is the aerodynamic advantage offered by travelling in an organised group. Geese are an obvious example.

Say you're a pink-footed goose wanting to go somewhere. You could set off by yourself, and you'd probably do OK. Your constitution is strong, and your body well suited to continuous flapping flight. It's a bit of a faff getting off the ground, for sure, because you're heavy for a bird, but once you're going you have plenty of stamina. While solo flight is plausible, there are great advantages to travelling in a group, and the trademark arrowhead formation employed by geese of all kinds is an effective illustration of this.

It's often likened to cycling in a peloton. The rider in front bears the brunt of the wind resistance, and the riders behind take advantage of the resulting 'hole' in the air. It's estimated that it's about 10–15 per cent easier to cycle at speed in another cyclist's wake. When they get tired, the leader peels off and the next in line takes over.

For geese it's slightly different. Flapping wings create both upwash and downwash. Upwash is good – it provides lift – while downwash is bad, producing drag. The trick is to take advantage of the upwash while avoiding the downwash, so the optimal position is behind the front goose and just to one side, your wingtip aligning with the wingtip in front. By adopting this position, the second goose takes advantage of the vortex created at the wingtip of the leading goose, the result of turbulent air emanating from the tip. They expend less energy, can flap less hard and less often, and can therefore travel further. Up to 70 per cent further, by some estimates.

To enhance the advantage of the upwash further, it can also help to synchronise your flapping with the front goose's. It's not exactly clear whether geese manage this simply by watching the goose in front, or whether there are more complex sensory systems at play. But they do it, adopting the V formation that will be familiar to anyone who lives on a migration route. It is of course hard work at the front, as it is for the leading cyclist in a group. So, after a while, the front goose takes a break and drops back, allowing one of its wing-geese to take over, and the journey is made that bit easier for the whole party.

A flock of geese, arrowhead flocks straggling across the sky, calling to each other with great braying honks, is a superb autumn spectacle, visible wherever they are found. But one species takes its migration flight to extremes.

The bar-headed goose isn't like other geese. It might look

similar – stout of body, long of neck, naturally belligerent of attitude – but one shouldn't judge by appearances. Internally, things are different. And those differences enable it to do something quite remarkable.

Head for Tibet, or Mongolia perhaps. Somewhere to the north of the Himalayas, in any case, and with lakes at high altitude. There you will find these birds, eating grass – as geese worldwide tend to do – and hanging around near their favoured rocky outcrops. They breed here, and when the time comes, off they set, rising as one from the water, a maelstrom of flapping and splashing and honking as they call to each other, communicating in ways unknowable to humans. Perhaps it's mutual reassurance ahead of a long and dangerous journey; perhaps they're just contact calls; perhaps they're just saying, 'Left, Steve! Left!'

They head south, aiming for their wintering grounds – a lowland swamp in India, perhaps. There's only one obstacle, and it is a significant one. To get to their destination they have to get past the Himalayas, the highest mountain range on the planet.

Other birds might take the long way round, skirting the Tibetan plateau to the east before taking a sharp right turn, the added length of the journey offset by the gains accrued by flying at low altitude. But bar-headed geese aren't other birds. They fly up, and up, and up. The highest recorded bar-headed goose was flying at 7,290 metres, and there are anecdotal reports of birds flying at 8,000. And so it is that these geese hold the altitude record, informal as it may be, and unratified by leading athletic authorities.

Why do they do this? Well, because they can.

Geese are continual flappers. They rarely resort to gliding, even when it might seem the most efficient thing to do – when descending, for example.*

Flapping flight is energy expensive, using between ten and fifteen times the amount of oxygen used at rest. And that's at sea level. The atmosphere at 6,000 metres and above isn't conducive to easy breathing, and particularly not if you're taking strenuous exercise. Up there the air is extremely thin, containing between a third and half the oxygen of sea-level air. The risk of hypoxia at that altitude is correspondingly greater. Not much fun, hypoxia – symptoms include fatigue, numbness, tingling, nausea, cyanosis, heart failure and death. And this is compounded by the difficulty of producing lift in thinner air. Even though birds are better at shifting oxygen around their body than mammals, those extremes would still cause severe problems for birds solely adapted to lowland life. As if that weren't enough, it's cold and dry up there, so, even factoring in the body heat produced by flapping flight, dehydration and heat loss are extra burdens to overcome.

To cope with all this, bar-headed geese have evolved specialist features. Compared with a species like the greylag, for example, their lungs are larger, they breathe more deeply, and their capacity for oxygen consumption is approximately twice as high. They are

* They do exhibit one remarkable behaviour, known as 'whiffling', in which the head and neck remain in their normal position while the body turns upside down. The result of this tinkering with the forces of lift is a rapid, zigzagging descent, which the goose will often quickly correct before coming in to land. It's suggested that they do this to avoid flying low over wildfowling areas, or as showing off, or when they spot a good feeding area and want to get down to it as quickly as possible without the tedium of a slow descent.

operating at the edges of what they can manage, for sure – their heart rate goes up close to its limit, but the benefits of the shorter journey must outweigh the disadvantages.

While they clearly relish a challenge, they're not masochists. They will drop to a lower altitude when the opportunity arises. And they will also take advantage of the phenomenon known as katabatic winds – winds formed by the movement of cold air down a slope – to make their journey a bit easier.

One possible explanation for this extreme behaviour is that when geese started making this journey, the Himalayas were lower. The choice between the high road and the low road would have been more straightforward, and the journey was passed on from generation to generation as 'the way to go'. As the mountains grew over time – a lot of time – the birds evolved the capacity to deal with the new challenges presented by the incremental increase.[1]

How birds know where to go is a source of fascination to anyone interested in migration. While our pink-footed geese – and plenty of other birds – travel in family groups, examples abound of juvenile birds making long migratory journeys for the first time (and only recently having learned how to fly) without expert guidance from an adult that can show them the way. Manx shearwaters, recently emerged from their birth burrows, are deserted by their parents and then make the journey from the west coast of Britain – an island such as Skomer, off the Pembrokeshire coast, home to about half the world's breeding population – to the east coast of South America. Or cuckoos, left alone by their parents before birth and getting, understandably under the circumstances, no help from their foster parents, heave to and fly from Britain to Central Africa, using nothing more than cussedness and an innate understanding of where to go. This seems to amount to a genetically coded instruction: go in this direction for this long and then stop. This does mean they're potentially vulnerable to going off course, simply by dint of setting out in the wrong direction without realising it.

Other cues used by migrating birds are similar to those used by migrating insects: they have a celestial compass, using the position of the sun and stars to guide them; once an area is familiar, they build a landmark map; and they react to changing day lengths. There is also the almost mystical concept of magnetoreception: the ability to sense Earth's magnetic field. It's been known for some time that birds possess this ability, but much of the subject, not least the exact mechanism or mechanisms by which it is achieved, remains so mysterious that it was described by Eric Warrant of Lund University in Sweden as 'the greatest Holy Grail in sensory biology'. There is evidence that magnetite in the upper mandible of birds' beaks plays a part,[2] and recent research suggests that particular proteins – cryptochromes – in a bird's eye are also involved.[3] This second phenomenon brings into play the quantum concept of 'radical pairs' – as always with anything including the word 'quantum', I would love to pretend I have the faintest idea what's going on, but in truth glazed eyes are never more than a sentence away.

The bar-headed goose's journey is a good advertisement for the benefits of honest endeavour, but for an alternative approach based on the canny harnessing of natural phenomena, we move from the Himalayas to southern Spain. Specifically, Punta de Tarifa – as close as Europe gets to Africa, the two separated by the Strait of Gibraltar. If you see a goose here, it will have strayed far from its intended path.

On a good day during autumn migration, there is an almost constant stream of birds of prey past the raptor viewing points at Punta de Tarifa – the numbers go up into the thousands. Booted eagles, black kites, griffon vultures, Egyptian vultures, sparrowhawks, marsh harriers, ospreys and plenty more. Pure heaven for birdwatchers, and the result of a happy confluence of geography and aerodynamic principles.

Take the booted eagle. It's a small eagle, as eagles go. About 50 cm long, with a wingspan of just less than triple that. Brown of plumage, bulky of chest, sharp of eye. Lives in woods, not averse to a mountainside, will eat your pet rabbit at the drop of a hat. This one might have come from Ghana, let's say. Somewhere south of the Sahara, in any case. And its destination might be Portugal. Not a mammoth migration in the grand scheme of things – 3,000 km or so – but quite enough to present dangers and challenges along the way. It has to cross the Sahara. It has to feed. It moves during the day and rests at night. Move, feed, rest, move, feed, rest. Repeat for between three and five weeks.

The booted eagle's deep, broad wings mean it's well adapted to what is known as 'static soaring'. Unlike geese, these birds are not especially strong flappers. Not feeble, but not strong. They will do it if necessary, but they are not light birds, and flapping uses a lot of energy, so they prefer to let the air do the work if they can. To do so, they exploit the specific properties of warm air over land.

Heat rises, so we're told. The physicists might quibble about this vague and oversimplified expression, but it will do for now. Perhaps it's more correct to say that heat moves from higher temperature areas to lower temperature areas – and when the sun heats the ground, this means the air moves upwards.

The sun shines. The ground gets hot. But some parts get hotter than others, and above those areas the air is, as you might expect, warmer. Up it goes, a column of warmth known as a thermal. At the top of the thermal, the air cools and descends, rejoining at the bottom, where the process begins again.

The canny broad-winged bird, looking to save energy, knows how to find the thermals, and it knows how to exploit them, spreading its wings and allowing the warm air to carry it up in a broad circle – in general, the larger the bird, the broader the circle, so smaller birds such as falcons might make it to the top more quickly. At the top, it glides towards the next column, often marked by cottonwool clouds formed by the condensation of cooling air. And here is where the larger birds regain the advantage over the smaller ones, by dint of their superior glide ratio.* This is the proportion of distance travelled for every metre of descent. My unaided glide ratio (and yours) is zero. If I were to put on a wingsuit – as used by skydivers and BASE jumpers – and jump from a plane,† it might go up to three. A monarch butterfly manages about six, while a flapling pterosaur's has been estimated at between ten and thirteen (the glide ratio varies according to speed).[4] The gliding crown goes to the frigatebird (we will encounter this bird again later), which manages up to 22 metres for every metre of descent.

When it comes to birds of prey, the champions – some of them with wings like barn doors – are vultures and eagles. Our booted

* It also goes by the rather more elegant name 'finesse'.

† The chances of my doing this voluntarily are, by an extraordinary coincidence, the same as my unaided glide ratio.

eagle, for example, while comparatively small, has a glide ratio of about ten. If it climbs to a height of a kilometre it can then glide 10 km, give or take, before having to deploy the flapping.

At its narrowest, the Strait of Gibraltar is 13 km. And as thermals don't form over water, this presents a small problem for the eagle. It must make sure it rises high enough before attempting the crossing. Even then, there will likely be a shortfall. And for the assembled birders, this is where the spectacle hots up. A lot of birds, unaided by helpful winds, find themselves running out of height as land heaves into view. This brings them closer to the watching hordes, who get a fine view of the birds as they flap landwards, sometimes with more than a hint of desperation.

The good news for the slack-a-bed birder is that you don't have to be up early to witness this spectacle. The formation of thermals depends on warmth, so the birds are slow to get going, waiting for the best conditions to develop towards the middle of the day. Sometimes they can be waiting a while. They're grounded by rain or low cloud, conditions that prevent the formation of thermals. Likewise heavy winds. And once over the water, they face another danger. The area is known for its unique conditions: between them, the Levante (easterly) and Poniente (westerly) provide more than 300 days of strong, steady winds. Beloved of windsurfers, but potentially lethal for migrating birds. The Levante can drift them out to the Atlantic, the Poniente over the Mediterranean. Either way, a splashdown will result in drowning, so they prefer, if possible, to wait for a lull.

For our booted eagle, like most such birds, the advantages of soaring flight over flapping are significant, with energy savings of up to 25 per cent. Any extra distance travelled to avoid flying over an expanse of water is well worth it. Contrast this with the bar-headed goose – different body shapes for different methods.

The booted eagle will find its breeding ground, breed, and then in September it will make the return journey, accompanied – or so we hope – by its offspring. These young birds – one or maybe two of them – will be making the journey for the first time. Once south of the strait, they might veer eastwards rather than westwards. Conditions are different in the autumn, and food sources may vary, so they employ a 'loop migration' strategy adopted by many birds – different routes in different seasons.

These are just a tiny sample of the millions of migrations undertaken every year. Birds on the move all over the place. Some make short hops; others undertake gruelling marathons. It might be as simple as the dispersal of newly fledged chicks away from their nesting site, or the more seasonal movement of their parents to and from the breeding place. It might be the belief-defying non-stop migration of an Arctic tern, or the more mundane movement of a ptarmigan from the top of a mountain to the bottom. Endless variations on a theme. For many, it means undertaking perilous journeys that often result in their death. And for all the journeys we know about, there are countless others presumed to be happening. Which leads to what might seem an obvious question.

How do we know?

The north German town of Klütz is not well known; 33 km north-east of Lübeck, it has a population of about 3,000, a manor house, a miniature railway, and a centre of literature named after novelist Uwe Johnson.

In 1822 a white stork was shot near Klütz. There is nothing unusual in this. Storks were, and are, common birds throughout mainland Europe, and hunting was a popular pastime. They're large birds, so eminently huntable. What was striking about this bird was that it already had a 76-cm metal-tipped wooden spear through its neck. Examination of the spear showed that it was of African origin.

You can't help but feel an upswell of sympathy for the stork. You get shot in Africa, fly all the way to north Germany with a spear sticking out of you, and no sooner have you arrived than you're shot again. At least the poor bird – now residing in the zoological collection at the University of Rostock – has its place in posterity. Much like our putative fossil in Chapter 8, it played an unwitting part in the advancement of human learning. In this case, this first 'Pfeilstorch' ('arrow-stork' – there have been more than twenty-five instances since) gave an answer to a question that had intrigued naturalists for centuries: where do birds go in the winter?

Some of the outlandish hypotheses concocted to explain the annual disappearance of birds from Europe now seem decidedly quaint. The idea of metamorphosis still held sway in some quarters. The most delicious version of this concerns the barnacle goose, which for centuries was thought to emerge from goose barnacles, those filter-feeding crustaceans that cling so stubbornly to rocks in intertidal zones. The black-and-white colouring common to the two creatures was the origin of the hypothesis. An alternative, or sometimes complementary, explanation was that the birds grew originally from trees overhanging water, falling off when ripe and assuming their adult form.

Gerald of Wales, writing in 1188 in *Topographia Hibernica*, went into extensive detail:

> There are likewise here many birds called barnacles, which nature produces in a wonderful manner, out of her ordinary course. They resemble the marsh-geese, but are smaller. Being at first gummy excrescences from pine-beams floating on the waters, and then enclosed in shells to secure their free growth, they hang by their beaks, like seaweeds attached to the timber. Being in progress of time well covered with feathers, they either fall into the water or take their flight in the free air, their nourishment and growth being supplied, from the juices of the wood in the sea-water. I have often seen with my own eyes more than a thousand minute embryos of birds of this species on the seashore, hanging from one piece of timber, covered with shells, and already formed. No eggs are laid by these birds after copulation, as is the case with birds in general; the hen never sits on eggs to hatch them; in no corner of the world are they seen either to pair or to build nests.

Centuries later, Gilbert White, the great eighteenth-century naturalist, engaged in vigorous debate about whether his beloved swallows spent their winters at the bottom of the local pond – an idea given credence by the testimony of fishermen who claimed to have hauled large numbers of the birds alive from under the ice. And a 1703 pamphlet, written 'By a Person of Learning and Piety', suggested birds flew to the moon.

It seems odd now that the subject was so little understood until relatively recently. Particularly when you read this, from good old Pliny the Elder, writing in the first century AD: 'The swallow . . . takes its departure during the winter months; but it only goes to neighbouring countries, seeking sunny retreats there on the mountain sides.' Go back another four hundred years or so, and Aristotle, while expounding the hibernation and metamorphosis ideas, also made the eminently sensible proposal that birds simply went somewhere else: 'Among birds, as it was previously remarked, the crane migrates from one extremity of Earth to the other . . . the phatta (wood pigeon) and the peleias (rock dove) leave us, and do not winter with us, nor does the turtle (turtle dove).'

Either those eighteenth-century naturalists should have read more Pliny and Aristotle, or they simply (and perhaps understandably) thought the idea of such tiny things flying such huge distances was too outlandish to warrant serious consideration.

Methodical recording of birds' movements took a step forward in the nineteenth century, although for a long time much of the information came from observers on the ground armed with binoculars – fine for monitoring low-flying species in good daytime visibility. These observations were supplemented by moon watching – counting birds as they pass in front of the moon at night. The results from these kinds of observation were useful, albeit with obvious limitations, given the numbers of birds that evaded detection – whether they were flying too high to be seen even with the most powerful binoculars, or were just too selfish to plot a path between the observer and the moon. The same applied to the monitoring of nocturnal flight calls, many of which remained undetectable until the development of sophisticated recording equipment in the second half of the twentieth century.

The idea of tracking birds with some sort of physical tag – preferably something less cumbersome for the bird than a spear – took hold at the turn of the twentieth century, and bird ringing ('banding' in America) quickly became widespread. Put a numbered metal ring on a bird's leg, release it, and hope it gets caught or found somewhere else so you can see how far it's travelled. The main problem with this is the sheer volume of birds that need to be ringed to yield meaningful results. It's one thing having a record of a bird where it was first ringed, but the most valuable information is generated when a bird is refound. This might be at another ringing station somewhere far away, the same place, or (quite often) when the bird is found dead. Approximately 2 per cent of British ringed birds are refound, so the results are understandably fragmentary. Nevertheless, the accumulation of records over the years enables researchers to build a coherent and consistent picture of what's going on out there.

One of the most eye-opening developments in migration research came with observations about the behaviour of caged birds. As the time for migration approaches, these birds show signs of agitation, especially towards dusk and during the night, when they would normally be asleep. They hop about on their perches, flutter their wings, and align themselves with their usual migratory direction. In 1966, researchers S. T. and J. T. Emlen designed a special funnel in the shape of an inverted cone. At the bottom, an ink pad; at the top, a wire screen to prevent escape. The birds' inky footprints showed the direction of their intended flight. These caged birds uniformly attempted to fly in the appropriate direction for that season's migration, and further experiments with magnetic fields helped establish the likelihood of magnetoreception in birds. As so often, the English expression describing the phenomenon – 'migratory restlessness' – seems clumsy beside the German word for it: *Zugunruhe*.

Radar technology was added to the armoury of migration trackers in the middle of the twentieth century – a by-product of its development in the war. Whether it's surveillance radar – excellent for general detection – or tracking radar – pencil beams for following individuals or flocks – this is not a cheap way of doing things, but, as with the tracking of painted lady migration, has proved invaluable in shedding more light on the unseen behaviours of birds, and in particular in learning about flight altitudes, speeds and other details that simply wouldn't be visible to even the most accomplished binocularist.

While ringing still provides the bulk of information about bird movements, all it can tell you is where a bird started and where it stopped. It doesn't tell you what happened in between. But now that gap is being filled. The biggest stride forward in the last few decades has been the development of satellite and radio technology, and in particular the miniaturisation of tracking devices. Now you can strap a GPS-enabled tracker to a bird, like a miniature backpack, and record its every move. This technology has revealed the true extent of some long-distance migrations, in some cases making already astonishing journeys seem scarcely believable.

Small birds will fatten up substantially in preparation for their migration. They need to be able to draw on every bit of energy in their long journeys, and while they might stop to feed, this adds time to the journey and leaves them open to predation. So they bulk up, sometimes reaching more than twice their normal body weight. But they can't just pile it on willy-nilly without thought for the consequences. The extra weight means they're slower and less manoeuvrable, and therefore a much easier (and more nourishing) target for any opportunistic predator they might meet on the way. But if they don't carry enough energy reserves, they risk dropping out of the sky with exhaustion before they reach their destination. They're after the Goldilocks scenario. In order to achieve this, they make other changes, too, growing their respiratory organs to cope with the energy-production needs of their journey, and shrinking other organs, such as the stomach and liver.

Many larger birds will carry less fuel and are able to travel faster. But this means they are less well suited to long periods of flapping flight, needing to stop more often to refuel. This, however, doesn't always apply. Birds nesting at high latitude – several species of goose, for example – might lay down extra reserves so they can lay eggs as soon as possible after their arrival at their breeding grounds. They arrive when vegetation is still sparse, and their breeding cycle is long, so no time is to be wasted. Females, with the responsibility for producing the eggs, lay down ample reserves before departure, and especially at their last stopover, increasing their weight by as much as 50 per cent so they can survive the rigours of egg production and incubation.

That level of weight gain is also associated with ultra-long-distance migrators. Take, for example, the bar-tailed godwit, a medium-sized (in the grand scheme of things, albeit large for a wader) shorebird with a long, very slightly upturned bill, and long wings. Its constitution makes it perfectly suited to non-stop flapping flight. It is able to increase its weight substantially without catastrophic restriction of its flight capabilities. It also shrinks its internal organs to save weight, in the manner of a smaller bird. Thus prepared, off it sets on a monumental migration. The exact starting point and destination vary, but typically they will start in the Arctic and travel to coastal regions of a similar longitude in the southern hemisphere. The record-breaking non-stop journey that caught the headlines in 2021 was undertaken by a tagged male bird (snappily named after its ring number, '4BBRW') that had already broken the existing record the previous year. The new mark was 13,035 km, from Alaska to New South Wales, and its non-stop flying time was 239 hours. That's an average speed of 54 km/h and just over 1,300 km a day. For reference, that's approximately the distance from Land's End to John o' Groats. Every day, without stopping.

The sheer stamina required for such a journey is one thing, but factor in the huge featureless expanse of the South Pacific, and you have a major feat of navigation too.

We're drawn to these extreme journeys. They're the glamour migrants. The companion world record to the bar-tailed godwit's extraordinary journey is the Arctic tern's. Dainty-looking little things, Arctic terns, like effete gulls. But appearances can be deceptive. These birds hold the record for the longest annual migration, travelling more than 70,000 km a year in a quest for eternal summer that carries them from Antarctica to Greenland and back again. In a twenty-five-year life, they might travel far enough to complete two round trips to the moon.

Migration is a continuum. Tempting and easy though it is to divide all birds into 'migratory' and 'non-migratory', nature enjoys defying our attempts to pigeonhole it. Those long-distance migrants – the godwit, the goose, the tern – capture the imagination, but for each of those gargantuan efforts, there are as many small-scale movements going on all the time. Five minutes spent on migrationatlas.org quickly turns into an hour or more as you're spurred to look up exactly how far and wide robins and blackbirds and blue tits and any number of familiar residents actually range (answer: much more than you'd think). Over a century of ringing records from European ringing schemes and the EURING databank has been consolidated with recent tracking data from Movebank to provide the most comprehensive summary of bird movement in Eurasia and Africa so far compiled. Thousands of thin lines, each one telling the story of a bird's journey, forming a network of resilience and endeavour.

However long the journey, it's possible only because the birds can fly. The immediate benefit of wings is the ability to escape from danger. You can fly up and away from those grasping claws and jaws. And then you can fly over the ravine to that ledge where the big bad thing can't get you. But you can also disperse far and wide, explore new places, exploit new resources. And you can move with the seasons in a way not available to earthbound species, whose journeys, while possible, are limited by the speed at which they can move.

Life is hard. But it might just be that bit easier if you have wings.

11

THE HUMMINGBIRD

Toronto, sometime in the 1990s. I'm in my mid-twenties.

It's been a while since I was really keen on birds. That was a childhood thing, given up when teenagerdom hit, bringing with it interest in Other Things. The passion I'd felt for them – an obsession, really – flared and subsided over a period of about six years, between the ages of eight and fourteen. A golden period, in many ways. But now I have moved on and am pursuing adult goals. Birds? I can take them or leave them. The latter, mostly.

Exceptions can be made.

My hosts are welcoming and kind, and they have a cosy kitchen where I can sit and drink coffee and recover from the effects of jet lag. Through the window, a feeding station. I watch idly, not concentrating. My brain registers something strange, can't work out what it is. Never mind. There is dozing to be done. I cradle my coffee, gazing without focus out of the window, vaguely in the direction of the feeders.

I realise what I find strange about the feeders: they seem to be filled not with seeds and worms and other such bird-appropriate food, but a pale liquid. Odd.

I look away, attention drifting.

What the hell was that?

It comes and goes so fast I'm pretty sure it's the product of a jet-lag-addled imagination. A flicker, then gone.

A few seconds later, confirmation that I'm not going mad. An actual visible thing, darting into view and staying there, motionless, apparently suspended by hope alone.

My first hummingbird.

Long-forgotten knowledge tickles the edge of my brain. Facts gleaned from the *Pears Cyclopaedia*.

The smallest bird in the world, at 5 cm long, is the bee hummingbird.

Hummingbirds can hover.

Hummingbirds can fly backwards.

Hummingbirds' wings beat at, like, a gazillion times a second.

Hummingbirds are amazing.

The vagueness of this remembered knowledge was, in a way, understandable. Hummingbirds weren't in my bird guides, for the simple reason that they are emphatically not European. In my mind they're birds of South America, birds of the rainforest, birds of a distant, unimaginable world. But here I am in suburban Toronto, casually watching the impossibly exotic, as close to them as to the familiar tits and finches and robins of the back garden at home.

The mystique surrounding hummingbirds is understandable. They resemble something you might come up with if asked to invent a miracle bird. Impossibly tiny, vividly colourful and aerodynamically improbable, they capture the imagination in ways that other birds don't.

It's impossible to think about hummingbirds without the words 'iridescent jewels' floating to the surface. Also 'whizz', 'whirr', 'dart' and 'fuck me, did you see that?' Now, as I watch these iridescent jewels whizz, whirr and dart back and forth from the feeder, I murmur, 'Fuck me, did you see that?', and understand what all the fuss is about.

As it becomes clear that I'm not going to stop watching them any time soon, another cup of coffee appears, then a pair of binoculars. I quiz my hosts. They know the birds only as 'the hummers'. Later I learn that these are ruby-throated hummingbirds (*Archilochus colubris*). About 8 cm long, weighing between 2.5 and 5 grams. Common birds, if you're in the right place.

But those bare facts are just the beginning of their story.

There is an illustration – you've probably seen it – of the evolution of humankind. Starting with a creature crawling from the primordial ooze and working from left to right, each image becomes progressively more upright, until the last one, which is a human (nearly always a male). It's been co-opted many times for humorous or satirical purposes, but it bugs me. It bugs me because it misses the point. Not only does that linear progression imply that each stage replaces the previous one (that's not how it works); it presents humans as evolution's pinnacle, its finest achievement, the natural culmination of evolutionary processes – this is, in a way, understandable, because as humans we place ourselves at the centre of everything. No doubt an aardvark, if pressed for an opinion, would regard the world from the point of view of the supremacy of aardvark-kind. It's a sin of omission – the illustration represents just one of millions and millions of strands. Every species that ever lived on the planet has its own version.

The penultimate image on a version for hummingbirds might be *Eocypselus rowei*, a fossil bird from the Green River Formation in Wyoming, about 50.6 million years ago. It's a good fossil, so good that even this layman, looking at photographs, can see that not only is it a bird but that the link with hummingbirds isn't outlandish. A head, spindly legs dangling down below a tangled skeleton, the blurred outline of wings, the arm bones clearly visible. If pressed, I might hazard a guess that this was the ancestor of finches, based on the clearly visible outline of its short, blunt beak.

Shows how much I know.

Eocypselus rowei was the last-known common ancestor of hummingbirds and swifts. Although it was neither a hoverer nor a dasher, its descendants would come to master both, each bossing the air in their own way.

For hummingbirds, this meant several physical adaptations, all of which are geared towards minimum weight and maximum flight capability. To lose weight, as all flying dinosaurs did, they lightened their bones, fused a good number of them and removed all unnecessary muscle and ligament. The muscles they do need – the flight muscles in the chest – became strong and proportionally large, taking up between a quarter and a third of their body weight.

And their keel, to which those muscles are attached, became correspondingly deep.

But the key to what makes a hummingbird a hummingbird is in the shoulder. Uniquely in birds, they have developed a ball-and-socket joint that allows their wings to rotate in a 140-degree arc. The practical upshot of this is that instead of doing what normal birds do – produce lift with a deep, angled downstroke and recover with as economical an upstroke as possible – hummingbirds produce lift on both strokes, their wings moving in a flat figure of eight, as if in imitation of insects. The angle of attack, though, is shallower – 15 degrees, on average, compared to 25–45 degrees in insects. The downstroke produces about 70 per cent of the lift, the upstroke providing the remaining 30 per cent. This almost continuous drive enables them to do the most difficult thing in all flying: they can hover. Moving slowly is hard enough – air flow over your flight surfaces is reduced, and you don't get as much help from it – but staying absolutely still is the hardest. Hovering – proper hovering, using only your own strength to suspend yourself in the air – requires enormous power and control, and every aspect of a hummingbird's body is adapted to deal with its rigours.

Their hearts are big – about twice the size, proportionally, of any other vertebrate – and they work relentlessly, their rate ranging from 250 beats per minute at rest up to 1,500 when in flight. This, combined with the efficient respiratory system common to all birds, means their capacity for oxygen diffusion is ten times that of other vertebrates of a similar size. And their metabolism works at a remarkable rate, about thirty times that of humans. This is the fastest metabolism of all vertebrates, which in turn means they live their lives on the edge of starvation, needing to replenish constantly to fuel their energy-hungry and hyperactive system. Over the course of the day they can consume up to six times their own body weight in nectar. To do so, they employ a remarkable miniature pumping mechanism that was only revealed on analysis of slow-motion video. Their tongues have tiny grooves on each side, which, when the tongue flicks out towards the food source, compress against the inside of the mouth. As the tongue touches the nectar, the grooves spring open, and the nectar is sucked in. Repeat fourteen times a second.

To counter this hyperactivity, they also have the ability to put their bodies into torpor at night or if food is scarce. Their heartbeat slows to a tenth of its usual active rate, with a corresponding drop in body temperature. This conservation of energy is especially important for small species living at high altitude, where the air is cool at night.

The nectar-drinking aspect of hummingbirds' behaviour is key to their success, and an apparent anomaly. Because while the genetic make-up of birds includes receptors for savoury tastes, it doesn't cater for the sweet. Instead, they have a mutation whereby the savoury receptors have adapted to detect sweet things as well. How exactly this came about isn't clear, but it's easy to envisage a scenario in which the extra energy provided by nectar is favoured by natural selection, leading to an increased taste for it, and eventually its dominance as the primary food source. From insectivore to nectar-drinker, in several thousand easy steps.

The flowers play their part in this relationship, providing sucrose-rich nectar to attract the hummers. Bees, which prefer fructose and glucose, steer clear. As hummingbirds use vision to find their food source, whereas bees use smell, the flowers colour themselves brightly – red and pink being the preferred shades – to ensure they get the right kind of visitor. The pay-off for the flowers is pollination. When the hummingbird plunges its beak into the flower to drink the nectar, it unwittingly picks up pollen on its forehead, spreading it around as it flits from flower to flower. That flitting, incidentally, isn't a random process. Their enlarged hippocampus means they can remember which flowers they've visited and, even better, which ones had the good stuff.

The co-evolution of flowers with hummingbirds means that many species are associated with one particular kind of flower, and their bills are shaped accordingly. Among the more extreme examples of this is the white-tipped sicklebill, whose bill curves dramatically downwards, the better to enable access to the nectar of Heliconia flowers. At the other end of the bill adaptation spectrum, the sword-billed hummingbird is the only bird in the world whose bill is longer than its body – most birds use their bills to preen, but for the sword-billed hummingbird this would be like a human picking their nose with a javelin. So it stands on one leg while performing the task with the other.

Those legs, incidentally, have shortened considerably since *Eocypselus*'s time – another weight-saving adaptation. Modern hummingbirds have tiny feet, just good enough for perching and occasionally scooting along a few centimetres, but not much use for anything else. Many birds need to push off with their legs to help them get off the ground but, for a hummingbird, strong feet and legs are unimportant. It's all about the wings. Its upper arm bone – the humerus – is shortened and, to help with stability and control, its finger bones are long and strong. When you're beating your wings at great speed (around 18 beats per second for larger species, and routinely up to 80 for the smaller ones), tiny adjustments can make a big difference.

The speed of hummingbirds' wingbeats is of course what creates the eponymous hum, by which it's possible – so I'm told by those who know – to identify which species has just whizzed into view. That's an impressive trick – almost as impressive as the spectacular courtship dives of the male Anna's hummingbird. The female sits perched, waiting to be wooed, while the male flies up to a height of about 30 metres.* It hovers for a few seconds at the top, like a gymnast summoning courage for a technically challenging routine. Then it launches into the dive, accelerating as it plunges towards the ground. Its aim is to impress the perched female – faster means stronger means better breeding stock. To this end it needs to accelerate as much as possible. The faster it's travelling as it approaches the bottom of the dive, the more impressive the key component of the display will be: the sound. As the dive bottoms out, the bird fans its tail feathers. This is the whole point of the exercise. The air thrums through the stiffened ends of the feathers, making them vibrate. This phenomenon is, I learn, known as 'aeroelastic flutter' – the self-feeding combination of aerodynamic forces and an object's own mode of vibration that, if uncontrolled, can lead aircraft or bridges to be destroyed. In this case, of course, the effect is more benign. The sound is like a sort of chirrup, lasting a fraction of a second and timed to sound as the bird whooshes past its potential mate. It's not dissimilar to the

* For perspective, that's the equivalent of a 1.75-metre human (me, for example) climbing to the top of Toronto's CN Tower. I know it's a false equivalence but I'm including it anyway.

sound the bird makes when calling, but because their sound box is so small, the tail-singing method is louder, and therefore a more effective advertisement of the bird's prowess. A diving male Anna's hummingbird has been measured at 20 metres (400 of its body lengths) per second. Dizzyingly fast, and generating G-forces well beyond what would cause even a seasoned fighter pilot to pass out.

When you consider hummingbirds, it's easy to be drawn – as I have – by the extremes, the record-breaking. Fastest, smallest, most aerobatic. They are a magnet for factoid fans. But while such things are a part of their fascination – how on earth do they do it? – I suspect there are other factors at play. It's not just their dominance of the air, mesmerising though that is. Beauty plays its part. Their delicacy, the iridescence of their feathers, the feeling that they are not quite of this world. There's something about them, too, that suggests they are on the verge of being something they are not. Hummingbirds and ostriches are at opposite ends of the avian spectrum, but they do have one thing in common: they are almost not birds. Evolution has taken them to the limit of what a bird can be. If you were to see a bee hummingbird – yes, I'm obsessing with the extremes, but it is the smallest dinosaur ever, so on this occasion I plead just cause – you might easily think it a kind of insect. Endemic to Cuba, the male of the species comes in at just 5 cm long and weighs about 2 grams. Using the traditional system of weighing birds, that is less than a quarter of the weight of a pound coin.

Given all this – their flying, their beauty, their absolute hummingbirdness – it's little wonder I'd kill to have them in Europe. But even if that will never happen, the tantalising fact remains that there were once European hummingbirds. When I say 'once', I mean approximately 30 million years ago, and I acknowledge that this, even by the standards of geological time, is not yesterday. But some of us have long memories and are still bitter about their decision to up sticks and start a new life in South America.

The evidence for hummingbirds' Eurasian origins is in two fossils, which, after sitting in a drawer in a museum in Stuttgart for years, were identified by Gerald Mayr in 2004 as ancestors of the modern birds. Long bills showed them to be adapted for nectar-sipping and shortened upper-arm bones pointed towards the ability to hover. They are, by some distance, the oldest such birds yet discovered, and they were, appropriately enough, named *Eurotrochilus inexpectatus* ('unexpected European hummingbird').

Not enough is known about *Eurotrochilus inexpectatus* to specify exactly when it lived – the latest approximation is 'sometime between 34 and 28 million years ago'. More precise is the dating of the first hummingbirds' arrival in South America 22.4 million years ago, although how and why they went there remains mysterious. One moment you have hummingbirds in Germany, and a mere ten million years later there they are in South America. Meanwhile – crucially – they have disappeared from their Old World hovering grounds, most likely never to return. They don't seem to have left relics anywhere else other than Europe, so the details of this mass relocation present quite the conundrum. While hummingbirds do migrate, non-stop trans-oceanic journeys are generally reckoned beyond them. The most plausible explanation seems to be that they went via the land bridge that then existed across the Bering Strait, and then diversified from there, although this also invites questions such as 'If they went that way, why aren't there any North American hummingbird fossils from that period?'

However you try to explain it, there remains a big gap in our understanding, but what we do know is that once they hit South America they diversified frenetically, radiating into nine families of about 350 species. By 10 million years ago they'd radiated into North America. Five or so million years after that, they made it to the Caribbean. Between them, they range from Alaska to Tierra del Fuego and live in habitats from rainforest to desert to mountaintop.

And while they were cornering the market in nectarivorous hovering, their cousins took a different path.

It is the most anticipated day of the year. Better than Christmas. Better than my birthday. Better than the first day of an Ashes series.

The swifts are back.

It's not all plain sailing. For at least ten days beforehand I risk crippling osteopathy bills as I crane my neck trying to catch that first glimpse, the sign that they've made the journey safely. And as I roam the streets of south London, I'm a liability. Never mind the modern curse of bumping into people glued to their phones. I'm the opposite – head up, staggering shambolically into bollards and bus shelters, ready to harangue random passers-by with a hoarse, deranged bark of 'They're here!'

Cause of death? Swifts. Or, more accurately, run over by a bus while looking for swifts.

When we talk about 'swifts', the default bird is the common swift – *Apus apus*. That scientific name, which translates as 'no-foot no-foot', gives you a clue to their preferred mode of transport. It comes from the long-held and genuine belief that swifts really did have no feet.*

While the common swift is the most abundant and widespread member of the family, ranging across Europe and Asia in the breeding season before wintering mostly below the African equator, the range covered by the 114 species of swift is in stark contrast to the relatively restricted distribution of the hummingbirds. Their cousins decamped to South America, but swifts radiated almost globally, each family favouring its own corner. Like the vast majority of animal species, they are absent from Antarctica, but all the other continents have some kind of swift somewhere or another, even if their presence is often in small geographical pockets. Some, like the São Tomé spinetail or the Javanese volcano swiftlet, are very localised, restricted to a single island, and sometimes similar species are distinguishable only by their specific geography and DNA. And while hummingbirds took the path away from *Eocypselus rowei* signposted 'iridescent hovering gems', swifts took a different fork. Their colouring, eschewing flashy iridescence, remained almost universally drab – dark grey, brown and black are the default shades. They might not have matched the hummingbirds' miniaturisation either, but they did remain relatively small – the largest swift is no more than 23 cm long, while the smallest members of the swiftlet family are about 9 cm. Their body shape – cigar or torpedo, according to preference – evolved to be more suitable for fast travel through the air than for hanging around flowers. Flowers very much aren't swifts' thing. They eat insects.

* I'm not sure where this came from, apart from 'back in the day'. Let's blame Pliny the Elder – usually a safe bet.

Lots of them. Not for them the fancy-shmancy bespoke bill shapes of hummingbirds. They don't want anything impeding the insects' journey into their mouths as they whizz through the air. So the bill is reduced to a little pointed nub – useful for snapping up a small fly or aphid or whatever poor unsuspecting flitter happens to be in their path. Meanwhile, the hind limbs, like the hummingbirds', have shrunk, to the extent that they have lost the ability even to perch. A swift's tiny curved toes are in fact deceptively strong, enabling them to cling to vertical surfaces, but their short, weak legs are ill-suited to the task of launching. While a healthy adult might manage fine, young or injured birds find it almost impossible to take off.

As for their skeletal anatomy, it is of course important for all flying animals to keep their weight down – but swifts don't need to take it to the same extremes as hummingbirds. The wings, instead of shortening, grew longer, broader, differently powerful, their muscular arc adapted not to hovering but to powerful flight. The elbow is close to the shoulder and rendered almost invisible by the long sweep of the wings to a tapering tip. The chord – the measurement from the front of wing to the back – is narrow, giving them a lower aspect ratio. Their chief asset is speed. Speed and manoeuvrability. Speed, manoeuvrability and a fanatical devotion to the air.

It seems barely credible that they can gather enough fuel to power their high-energy lifestyle, and indeed estimates of the numbers of insects they catch a day defy belief – up to 10,000, so I read. A life of constant activity.

Like hummingbirds, they offset this hyperactivity with the ability to send themselves into torpor, a strategy employed by nestbound swift chicks in periods of bad weather. A drop in temperature brings a decrease in insects, so the parents take a little mini-break to somewhere more benevolent – sometimes going as far as Germany in pursuit of better feeding conditions. This is normal enough. The extraordinary bit is that, like the migrating painted lady butterflies, they seem to leave before the weather arrives, staying ahead of it and returning when things have settled down. Meanwhile the chicks slow their metabolism and wait.

The life of the swift chick starts gently enough. After three weeks or so in the egg, they emerge blind and featherless – along with one or two siblings – into the loose nest of straw and saliva built by their parents. The parents feed them on wads of insects regurgitated from a pouch under the beak – up to 500 insects per wad. On this diet they grow and strengthen quickly. Within two months, the baby swift will be ready to fly. As the time approaches for them to leave the nest, they begin to get in shape for their maiden flight, practising push-ups on their wings and tails until they're strong enough. It's an endearing image. Two or three swift chicks, sitting in a tiny dark space, driven by genetic imperative to develop the muscles that will provide them with their livelihood.

The time comes. It's tempting to wonder – anthropomorphising just for a moment – if there is any doubt, any hesitation before they push themselves away from the nest and throw themselves on the mercy of their novice wings. But once it's out, it's out. There are no practice flights, no hesitation, no 'Hold on to me, Mum.' Out and off and don't look back. And then – and this is the bit that makes you blink – it will spend almost all of the next three years in the air.

Three years.

This fact always elicits a raised eyebrow, so convinced are we that the ground is where it's at. It's reflected in our language.

Someone who is grounded or down to earth is solid and sensible, not like those airy-fairy types with their heads in the clouds, living in cloud cuckoo land and prone to pie-in-the-sky ideas. But that's because we're not swifts. The air is their medium, as a fish's is water and an Englishman's is mild embarrassment. They do everything in it, leaving it only to nest and raise their young. And the young swift has no need for that yet, so in the air it stays.

Even people with no interest in birds might make an exception for swifts. Their magic captures the imagination. Part of the appeal lies in their vocalisations. It's so easy to use the word 'scream' or 'screech' to describe those high-pitched sounds, but those words carry overtones of horror, fear, ugly shrillness, tapping into their old folk name: 'devil bird'. Slow them down and they sound like a conversation.

Squeal? Is squeal better? Better, perhaps, at encapsulating the sense of excitement the sound inspires in the human heart. This human, at least. There is a particular thrill to be had watching them from a second-floor balcony as they give their performance. And it does feel like a performance sometimes, as if (more anthropomorphising, I'm afraid) they know we're watching. When they fly past in formation, writhing in and out of each other's slipstreams, they draw the eye and keep it with them, tracking their every move as they bank and skid and whoosh. The temptation to dust off comparisons with *Top Gun* is almost overwhelming. Then they swoop in towards you – six or seven of them flying in close formation. Some will peel off at the last second, dart up under the eaves and grapple their way into the confined space they call a nest. Sometimes they merely cling for the briefest moment to the wall before swinging down and away for another go-around. Avian parkour of the highest order.* And then they pull up steeply and go into cruising mode, describing broad arcs in the sky, gliding up and around, coming together in a group to flash past the window with their trademark *squee*, which I always – and completely unscientifically – choose to interpret as sheer delight at the fun of it all. A dispersal, each tracing its own path, gaining height until they're no more than a loose group of squiggling parentheses against clear blue sky. A wing waggle, crest an invisible rise, hang motionless in the air, then flip and turn and back and round again.

They continue for a while, screaming around at roof height, until they complete their last fly-past of the day against the twilit sky. Then they take it to the next level, above the trees, where they hawk for insects for as long as they can, circling higher and higher, soaring, up, up, up, beyond our sight and into another realm.

Hummingbirds and swifts redefine what it is to be a bird, taking the aerial life to its limits. Their abilities are so far removed from the flightlessness of the ratites that it's scarcely believable they belong to the same class. But there's more than one way to dominate the air, and for some birds flapping is too much like hard work.

* Germans and Danes, using their imagination more effectively than those charged with giving swifts their English name, call them 'wall sailer'.

12

THE ALBATROSS

There is something thrilling about cliffs. This is where worlds meet. Part of the appeal is that you're looking out across the water, its wide expanse a reminder of our smallness. And sight of the sea brings on thoughts of long journeys, adventures. Never mind that I don't go on adventures – it's enough to live vicariously, imagining the adventures of others.

Most of the world's birds live solely on land, seldom encountering a body of water bigger than a bird bath. For some the sea is a barrier, either to be avoided or to be crossed as quickly as possible twice a year on migration. Others take advantage of the benefits of life near the water – the easy availability of invertebrates on marshes and shorelines, for example – without committing fully to it. And then there are those for which water is a second home. They count among their number some of the most spectacular flyers on the planet.

I don't often travel far to see birds. But sometimes you just happen to be in the area when something interesting or rare turns up, and . . . well, you know how it is. In this case, the something interesting or rare shouldn't be anywhere near Yorkshire. It should be hanging out with its mates somewhere in the southern hemisphere. The Falklands, perhaps, which holds most of the world's breeding population. Not Yorkshire.

It is a black-browed albatross, one of the group of birds also known, entirely pleasingly, as 'mollymawks'. Black-browed albatrosses, like nearly all the species of their family, are overwhelmingly birds of the southern hemisphere. And while they are renowned for their nomadic tendencies, the North Sea is quite the peregrination. But it happens sometimes. The suggestion is that a storm might have carried this bird north of the equator, dumping it on the wrong side of the low wind trough known as 'the doldrums', and it's been unable to find its way back.

Of all the albatrosses, the black-browed is the one most likely to be found in the northern hemisphere, but it is still a sighting of some note. He – and they're pretty sure it is a he – has been hanging around these cliffs on the Yorkshire coast for several weeks now. It is, they think, his fourth visit to the same place. He arrives every year within a few days of the same date and stays for the summer, dividing his time between the cliffs and the open sea. Sometimes he nips down the coast to Flamborough Head for a change of scenery, but mostly he gravitates towards the 100-metre chalk cliffs of RSPB Bempton.

Does he somehow sense that not everything is right? Does he scan the gannets and gulls, looking for a bird in his own image, wondering why all the others are so small, their wingspans so comparatively puny? Does he realise he's in the wrong hemisphere?

Albie, they call him. And today, presumably because he knew I was coming, there's neither hide nor feather of him.

It's going to be a long day.

Samuel Taylor Coleridge has a lot to answer for. The image many people have of an albatross, whether or not they've read *The Rime of the Ancient Mariner*, is as a burden, something to be carried in perpetuity as punishment for one's sins. It's an image recalled with absurdist humour in the Monty Python 'Albatross' sketch, with John Cleese as a furious ice-cream girl charged with the task of selling a dead seabird to bewildered punters who only want a couple of choc ices.

Striking as those images are – both featuring, as they do, birds that have ceased to be – it is much more appealing to think of these birds as living, breathing things, ranging far and wide across the southern oceans, embodying freedom of spirit. All birds are, to an extent, unknowable, although prolonged study of their lifestyles and behaviour, along with a hefty dose of deduction, can give us an idea of what makes them tick. Seabirds, living much of their lives away from the human gaze, retain an aura of mystery. And albatrosses take the prize. We encounter them relatively rarely. They live a pelagic lifestyle, coming to land only to breed, and even then choosing remote islands, wisely eschewing human company at all costs. Albatrosses thrive in conditions that would make most humans cling wanly to the ship's railing while trying to keep their lunch down. Given a favourable wind and an open ocean, they are about as good at being in the air as anything else on the planet.

The albatross is the largest member of the order Procellariiformes, which also includes petrels, storm petrels, shearwaters and diving petrels. The exact organisation of the relationships in this group is – almost inevitably – uncertain, and the subject of continued debate and revision, but it seems that they all evolved from a smaller bird not dissimilar to today's storm petrels. The oldest confirmed albatross fossil – *Tydea septentrionalis* – is from Belgium in the Oligocene, about 30 million years ago. This bird seems to have been similar in size and shape to our friend the black-browed albatross, a sign that albatrosses had already diversified from their smaller ancestors and adopted the distinctive aerial technique that sets them apart from other birds.

The word 'flying' is almost inappropriate in the context of albatrosses, because they spend comparatively little time engaged in powered flight. Their wings are long and slender, giving them a very high lift-to-drag ratio (19:1, compared with, for example, the swift's at 10:1 and the starling's at 5:1), and an equally impressive aspect ratio of up to 15.

What those numbers amount to is that they are really good at soaring. Specifically the kind known as 'dynamic soaring' – distinct from the 'static soaring' adopted by the migrating birds of prey using those rising thermals – with which they can stay in the air for days on end with barely a flap of the wings.

It goes something like this.

The open ocean is often rough, with strong winds creating sizeable waves. The wind speed near the water's surface is low, increasing gradually the higher you go above it. Albatrosses exploit the predictable variety of wind speeds in this boundary layer – from the surface up to about 20 metres above the water – to astonishingly good effect. This exploitation of the phenomenon known as 'ground effect' is also used, in different circumstances, by birds such as cormorants, which can often be seen flying low and economically over rivers, using the metre or so of air between their body and the water's surface as a buoyancy aid. It works over land, too, but is more common over water, which is generally less likely to throw up random obstacles such as rocks, trees or Empire State Buildings.

The albatross flies into the wind, which gives it lift but also slows it down. Just as it's about to stall, it turns 180 degrees and dives down into the valleys between the waves – into the slower wind. Once it hits maximum speed it picks its moment to turn into the wind again, its momentum helping it wheel up above the waves into the faster air, which gives it lift but also slows it down. Just as it's about to stall, it turns 180 degrees . . . Repeat. Repeat. Repeat. With this zigzag pattern it makes its way across the ocean's expanse, making corrections with small adjustments of the angle of the wings, and flapping only when absolutely necessary. Streamlined wings, with nice thick leading edges and a good taper to the trailing edge, enhance the technique's efficiency, keeping drag to a minimum. The length of the wings helps too. Lift is weakest at the wingtip, so you want those tips to make up as little of the wingspan as possible.

Holding your wings out stiffly for hours on end might lead to fatigue, but albatrosses are equipped with a handy catch mechanism in their shoulder joint – a sheet of tendon that extends to lock the joint in place, saving valuable muscular energy.

Wind direction plays a large role in this endeavour. Any flying thing prefers to take off into a head wind – sometimes albatrosses merely have to open their wings and let the wind do the lifting for them – but flying into one is far too much like hard work. For trans-oceanic soaring, a tail wind is what you need. With the help of a tail wind, the albatross's flight speed nearly doubles, while its heart rate remains approximately the same as when at rest.

Even when the seas are stiller, albatrosses can manage. All they need are waves, so they can adopt the technique known as 'slope soaring'. Air is pushed upwards as it meets the waves – much in the same way as updrafts occur at cliff edges or along dunes – and they can take advantage of the rising air to carry them to the next one. It isn't as efficient as dynamic soaring, but it'll do until a proper wind comes along.

The compromise – there's always a compromise – is that they are comparatively weak flapping flyers, and in particular not good at flapping for extended periods. Which works out OK, because they don't often need to.

Given the potentially vast expanses of the ocean, soaring seems a good strategy to adopt, and other seabirds, while not quite attaining the expertise of albatrosses, use it to great effect. Gannets, with long, slender wings narrowing to a tip, are well suited to it, and gulls too are not averse to an energy-saving glide when the need arises – which is quite often. But the soaring prize goes to birds built to a quite different body plan. Frigatebirds (also known as 'man-o'-war birds'), with a combination of highly pneumatised bones and large wing-surface area, have the lowest wing loading (surface area divided by weight, in case, like me, you forget these things all too easily) of any bird. Combined with a high aspect ratio, this places them in a special section all by themselves – they're both superb soarers and very manoeuvrable. Their soaring skills are such that they can stay in the air for extremely long periods – one satellite-tagged bird managed two months. This is particularly useful because, unusually for seabirds, their plumage is extremely susceptible to waterlogging – to the extent that, if it becomes too saturated, they drown. Strange that a bird which spends so much of its time above water should be so averse to it. Like swifts, they're capable of sleeping on the wing, one half of their brain staying awake to stop the bird falling out of the sky while the other half takes a nap.

There are no frigatebirds at Bempton Cliffs as I stand at the viewing point. They're birds of tropical waters, and if one were to appear it would be an even bigger deal than the albatross – just two have ever been seen in the UK. As I scan the sea below, there's also still no sign of Albie the albatross.

Despite Albie's absence, there's plenty to enjoy. A well-populated cliff is one of the best places to watch flight at its most thrilling. Here, in the right season, are seabirds galore. Half a million of them, so they reckon. Gannets, puffins, guillemots, razorbills, fulmars, kittiwakes, gulls and shags. And if that's not enough, you can turn round and watch a skylark climbing into the sky above the surrounding farmland, or head back along the cliff to commune with the thriving tree-sparrow community. Something for everyone. My mother's warning words – 'Don't go near the edge!' – ring in my ears from across the decades as I look down (with a hint of vertigo) on the cliffs and sea below. There are birds everywhere, clinging to the sheer cliff face, wheeling round in lazy arcs, flying with purpose across the water. So many of them, each on its own personal mission, the whole resembling a scene conjured by a random movement generator.

The contrast in flight styles is stark. Gulls and gannets exude a leisurely vibe as they float past on outstretched wings. Far below, whirring bullets whizz back and forth, tracing criss-cross paths over the water's surface like a fiendishly complex cat's cradle. These birds – puffins, guillemots and razorbills – are members of the alcid family, and offer as dumpy a counterpoint to the broad-winged elegance of the albatrosses as you can imagine in a seabird.

The puffins, with their multi-coloured bills, catch the eye. Their short, compact bodies are ill suited to soaring, but their abilities extend into areas albatrosses can only dream of. In flight they exude a manic energy, as if they're desperate not to get their feathers wet. But they're happy in water. They dive to feed, and need to be fast and mobile to catch the sand eels and other fish that squiggle around nearish the surface. As water is approximately 800 times denser than air, the adaptation required to propel yourself effectively in both mediums is specialised. The costs of flight for puffins and their relatives are high. They can get away with doing both only by staying relatively small, and even then they are pushing the limits for both activities.

But if their wings seem more adapted to propelling them through the water than through the air, this is slightly misleading. All you need to do is watch them coming in to land to see that they possess important manoeuvring skills. Winds can be strong here, whooshing in from the sea and hitting the cliffs to create updrafts, swirling vortices and unpredictable eddies. Yet puffins plonk themselves down on a handkerchief-sized piece of land outside their burrows, calibrating their approach with judicious use of wing brakes, and waddling off nonchalantly after landing, as if it was all no big deal.

A similar feat of control is executed by a gannet just a few yards away from me. Thrilling birds, gannets. The long, slender wings, tapering to black-tipped points; the peachy shades on head and neck; the dagger bill. This one glides easily towards the cliff, exploiting the updraft, its composure only briefly disturbed as it prepares for landing on the cliff edge. Here the length of its wings works ever so slightly against it, although it does its best to style it out. It allows the wind to lift it to the cliff's edge, surfing the air for a few seconds, making minute adjustments with wings and tail, the wind ruffling the feathers on the trailing edges. Only as the landing becomes imminent does a hint of panic enter proceedings. Tail fanned, wings angled, webbed feet braced against the air to help – every bit counts. Drop, fold wings, settle, relax.

An albatross might manage the same nonchalance when landing on a ledge, but an equally likely outcome on flat ground is an undignified crash landing – a disadvantage of their soaring-adapted wings. They have a limited capacity to vary the surface area or camber of their wings – a common and effective braking strategy for birds – so on approach they just lower their feet and start pedalling in the air. If landing on the water, they can just splash down feet first. But touchdown on land requires a bit more control – they're often travelling faster than they can run, so the landing ends in an ignominious tumble, like someone misjudging the pace as they get off an escalator. To my eyes this apparently undignified spectacle has a strange nobility – they're making the best of a bad lot, accepting that landing will never be one of their strengths, but secure in the knowledge that they shine in other areas.

One of those areas is scent. Albatrosses are equipped with very large and effective olfactory bulbs, so their sense of smell is superb, and they adopt a similar approach to fruit flies when tracing scent, zigzagging to find a plume and then following it to the source. What they're sniffing for are fishy odours – chemicals released by the squid, fish, fish eggs, krill and other crustaceans that form the bulk of their diet. Oceans are big, and food is unevenly distributed, so the ability to lock on to a trail leading you to a meal is invaluable. Their cousins in the group of seabirds known as tubenoses – petrels, fulmars, prions and shearwaters – adopt a similar approach, although their slightly different diet means they hunt for different chemicals. But the main difference between their strategies is that albatrosses enhance their search with visual cues, whereas the other tubenoses hunt by smell alone. It's thought that the reason for this variation is related to their different breeding environments. Albatrosses are the only members of the family to eschew burrow-nesting, raising their single chick above ground where they can get to know their parents by sight and smell. A petrel chick spends the first weeks of its life underground, where scent is the only guide.

While the tubenoses are united by the prominent facial feature implied by their name, the specifics vary. Petrels have a single tube on top of the bill, while albatrosses have twin tubes running along the upper sides. The function of these tubes is threefold. There's the scent, already mentioned. They also help them gauge airspeed as they fly, enabling the search for areas where dynamic soaring will be more effective. And at the back of the nostrils, just above the eyes, special glands help them excrete salt so they can drink seawater – a handy trick if you spend a lot of time at sea.

Finding food is one thing – catching it is another. The feeding habits of all waterbirds are closely allied not only to their ability in the air but how comfortable they are immersing themselves in the water – whether at sea or inland.

Albatrosses are predominantly surface feeders, sitting on the water and picking off what they can reach. Others prefer to exploit resources deeper down, and for this they need to dive. Plenty of birds manage this from the surface. Cormorants are a good example, often giving a little jump before diving powerfully, their webbed feet bearing the brunt of propulsion with a little help from the wings. Their strength underwater owes a lot to their plumage, which (like that of the frigatebirds) is less waterproofed than many birds'. Long thought to be the result of a lack of oils in the feathers, this is now known to be the result of their microscopic structure. The wettability of their plumage means it traps less air, so the birds have reduced natural buoyancy and expend less energy in the dive. On returning to the surface, they will typically find their way to a perch and hang their wings out stiffly. Intuitively, this behaviour seems obviously designed to help dry the feathers, and some studies have borne this out, but in some species it's thought to have the added benefit of thermoregulation.

While cormorants are able to use their waterloggable plumage to their advantage, frigatebirds must avoid contact with water completely, so they take a different approach, flying close to the water and scooping up small fish and suchlike that have been driven to the surface by predators such as tuna. But they are also notorious kleptoparasites, using their manoeuvrability to harry other birds – gulls, terns, shearwaters, whatever they can find – into regurgitating or dropping their prey, which they then scoop up in mid-air. This is a tactic also employed by skuas (known outside the UK as 'jaegers') – chunky, powerful, furious-looking birds that use their strength and manoeuvrability to bully their victims even more ferociously.

Relying on food that lives close to the surface, but adopting a different strategy, are terns. Light and slender of body, flying with elegant and controlled movements, they are like refined gulls. They hover a few metres above the water, pointed wingtips describing a nifty figure-of-eight pattern – they are sometimes known as 'the swallows of the sea' – before plunging down to and below the surface and coming up with a fish. The rapier can be as effective as the bludgeon.

Or you could use a flick knife, like the skimmer family. These birds give frigatebirds a run for their money in the category 'peculiarly distinctive waterbirds'; they are also known as 'scissorbill' or 'cutwater', names that will give you an idea about their feeding technique. The construction of their bills is remarkable, the lower mandible markedly longer than the upper, the whole thin and sharp. Long-winged, flying steadily with deceptively languid beats, they move low to the water, slicing the surface with the razor-thin lower mandible. When it encounters a fish, the upper mandible snaps down to ensure the catch. Uniquely in birds, their pupils are vertical slits, allowing them to keep the bill within their field of vision and make sure it's best placed to scoop up the small surface fish that make up the bulk of their diet.

Rich though the pickings might be near the water's surface, there are resources to be exploited further down. All you need is the ability to get there, and while diving from the surface is one way, you can also use gravity and speed to propel you to sometimes surprising depths. The most spectacular exponents of this technique are gannets. Those slender wings, with their wide span, might give them a superficial resemblance to albatrosses, but not for them the fiddling around on the surface waiting for an obliging squid to come into their purview. They're built for diving. They fly languidly up to the optimum height – maybe as high as 30 metres – execute a little banked roll, and enter the dive, their streamlined bodies slicing through the air as they plunge towards the water, wings retracted and pressed into their sides. At the last minute they tuck their wings even closer to their body, close their nostrils, and bring special second eyelids – nictitating membranes – down over their eyeballs. Their speed on entry is up to 100 km/h, and they can reach depths of 11 metres, swimming deeper from there in pursuit of their prey – fish small, medium or large – which they nudge downwards before catching and swallowing whole. They minimise the effect of their impact by entering the water at a slight angle, and the configuration of muscles in their neck and head helps them avoid injury.

Pelicans use a similar technique. They have the same adaptations – nictitating membrane, muscle network, swept-back wings just before impact – but the difference is that they're after fish just below the surface. So they employ emergency airbags to arrest their descent – the airbags take the form not just of the sacs that all birds have as part of their respiratory system, but also the prominent expandable pouch – featherless skin reinforced with collagen – which hangs down from their long, streamlined bill and doubles as a place to keep their catch.

But if you're after a display of daring to put the albatross's fiddly surface-picking to shame, head for a freshwater lake somewhere near an osprey's nest. There is a beaky nobility to a perched osprey, head held high, yellow eye prominent, tufted nape giving the impression it needs a haircut. And there are few more spectacular sights than a hunting osprey in full cry. A heron catching a fish, stalking its prey patiently along the shoreline before spearing it with its dagger bill – that's one thing, and a fine display to watch. But an osprey hurtling towards the water with folded wings is a different kind of spectacular.

The approach is often shallow. Shortly before impact, the osprey stretches its feet out in front of it, ready to grab its prey. The hit is messy, water spraying everywhere. And then it gets difficult. Unless it's managed to grab the fish from the surface and then fly clear without submerging its body, the osprey is now wet. Hauling yourself out of water requires a great deal of energy, especially when you've just voluntarily drenched yourself. The osprey must produce enough lift to break clear of the water's pull with a severely reduced flapping angle.

It strains, its long wings – three times its body length – slapping heavily against the water's surface. Somehow it manages to haul itself clear enough to start a normal stroke pattern, its legs pumping to help with the momentum while making sure the fish remains clasped in those sharp talons.

If saturated, the osprey might shake itself like a dog to get rid of excess water. And now it does something very clever. Partially waterlogged, carrying a heavy fish, and unable to tuck its feet up to its body, the osprey is subject to much more drag than is ideal. It adjusts its grip on the fish, making sure it's facing forward rather than broadside into the wind. Off it goes, its burden slightly lessened by its instinct for drag reduction.

Marginal gains – as important in the natural world as they are in professional cycling.

The wind drops for a minute. A herring gull comes into view, soaring gently, white against the clear blue sky, backlit by the sun. It seems to wait until it knows I'm looking. Then it floats almost to a standstill and, without appearing to move a feather or a muscle or any bit of its anatomy, wheels round and goes back the other way. It's the merest tilt, a slight shift of angle against the air, no more than five degrees. It holds its wings stiff and still and just lets physics do its thing, and within five seconds it's turned 180 degrees and has floated away out of my sight. I get the distinct impression it did it only to show me it could.

It knows, dammit. It knows how to manipulate the air, how to move in it and through it, to save all its energy for a good old squawk. It does it, easy as winking, leaving me to ponder the relationship between air, water and land, and the nature of these alien lives that intersect with ours. Pity the poor herring gull, reviled as a chip-nicker, when all it's doing is being a herring gull – opportunist, canny, more than happy to take advantage of humans dim enough to eat their food within view of an agile scavenger. I wonder how we have got to the point where, instead of watching it with silent admiration for its resourcefulness and pluck as it manipulates the air to its own ends, we'd rather shout at it for taking advantage of our own failings.

The gull floats out of view behind me. I turn round, and there he is, floating in on stiffly held, gargantuan wings. There's no danger of mistaking Albie for a gannet or a gull or a kittiwake. He is a beast of a different hue. Even from a distance the proportion of his wings draws the eye, the length of them, their slenderness. As I get him in the binoculars and trace his path, he comes close enough for me to see that prominent bill, the hook at the end, the large black eye. He pulls off the neat trick of combining size with delicacy.

What an animal. You could swear he knows he's the centre of attention. Owning the stage, you might call it.

He soars up towards the cliff, circles once at a shallow angle, then approaches it head on, allows the updraft to lift him and place him down on his ledge, and there he settles, folding his huge wings into his body, apparently oblivious to the continued activity of the smaller birds around him.

And relax.

I keep thinking about his journey. It's impossible not to. Albatrosses are natural wanderers. They live long lives, and while they travel great distances in their lifetimes, they are loyal to both breeding site and partner, returning to the same place. They lay one egg every year or two, taking it in turns to incubate and then look after the chick, while the partner goes on immense foraging journeys, ranging over thousands of square kilometres of featureless ocean. Over the course of their lives, it's estimated that an albatross can travel nearly 5 million kilometres. Using the conventional currency for such things, that's six times to the moon and back.

It exhausts me just thinking about it. But seeing a bird so far from its natural home adds an extra dimension, taps into human feelings of isolation, the single misplaced bird among thousands speaking to anyone who has ever felt alone in a crowd.

But while this albatross is memorable, not just as an admirable exponent of one of the extremes of aerial competence, but as a rare encounter with one of the world's more majestic flyers, there are examples of everyday brilliance much closer to home. Sitting right smack in the middle of the incredible spectrum of flight are birds of such ubiquity and familiarity that they've become all too easy to ignore. Just step out of your front door.

13

THE PIGEON

My local high street. Average south London. Shops, traffic, people. An urban environment, much like many others.

You'd have to be looking fairly hard to find wildlife here. Hotbed of the natural world, it isn't. But there's one exception. They stalk the pavement outside the bakery, pecking for scraps; they mooch around the entrance to the cemetery, leaving it to the very last second to escape death at the wheels of a hearse; on the station platform, a male pursues an unwilling female, neck feathers puffed up in an apparently doomed effort to make her fancy him.

Pigeons.

They are everywhere, colonising our urban areas, breeding year-round and making themselves comfortable in the ledges and crevices that abound in such places. Many regard them as a scourge. Unsanitary. Rats with wings. While there's no doubt that too much of any species can be off-putting, I humbly suggest a moderate reassessment. Because pigeons are amazing. And while they are, on the face of it, less eye-catching than those aerobatic geniuses hummingbirds and swifts, they are strong and capable in the air, and also stand as a proxy for the many generalist flyers in the avian world, each of them achieving extraordinary feats of aerial agility on a daily basis.

We might not appreciate them, but one creature here is grateful for their abundance. It sits high up in the church tower, small but clearly visible if you know it's there. It's been there all morning, doing nothing more than watching the world go by and giving itself the occasional preen.

Peregrine falcon.

They say that if you see a bird of prey, it has long since seen you. The long-distance vision of peregrines is about eight times better than ours, and their flicker fusion threshold – the rate at which they perceive flickering light as a steady image – twice as good.[1] A peregrine could see a pigeon a mile off. But it doesn't need to. There are plenty much closer.

The pigeons – plump-breasted, tasty pigeons – get on with their lives below, blissfully unaware that one of them will shortly be lunch.

The fossil record of the pigeon family is sketchy. You might even say meagre. Their close relatives the sandgrouse are represented by four species in France from between 38 and 20 million years ago, but it's not until at least the early Miocene (23 to 16 million years ago) that members of the Columbiformes (to give them their formal name)

make an appearance. One such bird is the one called *Rupephaps taketake*, a large fruit pigeon from 19 to 16 million years ago, known from a single bone found in the St Bathans Fauna in New Zealand.[2] On scraps such as these a picture of the past, however faint, is built.

While they're not well represented in the fossil record, there's no shortage of living pigeons. There are 348 species worldwide, and they're found just about everywhere except in the most extreme of environments. They're all built to a similar body plan, but the range in size is large. The smallest, about the size of a sparrow, is the dwarf fruit dove; the largest are the four species of crowned pigeon, which would give our bar-headed goose a run for its money. But the bird under scrutiny here is the feral pigeon – *Columba livia domestica*. A middle-of-the-range pigeon – the Volkswagen Golf of birds – rampantly available under a railway bridge near you.

It was Charles Darwin who suggested that domesticated pigeon breeds, in all their enormous diversity, were all descended from the rock dove. This wild bird is now comparatively rare, but still holds a small population on cliffs around the Scottish and Irish coasts. We started domesticating them at least 5,000 years ago, and our tinkering with their genes has resulted in wide variation. Selective breeding has brought us pigeons specialised for flying-based tasks such as racing and message carrying, but others fall into the 'fancy pigeon' category – birds bred to favour certain traits, such as colour, plumage features, behaviour and vocal characteristics. This discipline has brought us fantails, pouters, jacobins, highflyers, frill-backs, trumpeters and many more – at least 1,000 of them within the single species *Columba livia domestica*. Inevitably, domesticated birds escape, and those birds formed the basis of the abundance of feral pigeons that we love to hate in our towns and cities.

Darwin's interest in them is unsurprising. He bred several varieties at his home in Kent, and not only do they feature in *On the Origin of Species*, but also merit an extended passage in *The Variation of Animals and Plants Under Domestication*, with observations on their wing anatomy and flying habits. Of the breeds under his scrutiny, the tumbler pigeons stand out for their flying exploits. Darwin included observations on them by one of his correspondents, a pigeon fancier and breeder by the name of Brent:

> Some fly round with the flock, throwing a clean summersault every few yards, till they are obliged to settle from giddiness and exhaustion. These are called Air Tumblers, and they commonly throw from twenty to thirty summersaults in a minute, each clear and clean. I have one red cock that I have on two or three occasions timed by my watch, and counted forty summersaults in the minute. Others tumble differently. At first they throw a single summersault, then it is double, till it becomes a continuous roll, which puts an end to flying, for if they fly a few yards over they go, and roll till they reach the ground.

This extraordinary demonstration of inbred aerial gymnastics is by no means the norm for this species, but whichever variety you choose – excepting the disproportionately large ones bred for the table – pigeons are excellent flyers. That plump breast houses a low-slung and proportionally huge keel; they have a short, stocky upper arm, and the wings are both strong and aerodynamically shaped. Their wingspan is neither extravagantly wide nor comically tiny, and the surface area of the wings is well in proportion to the substantial amount of lift they can generate with their chunky flight muscles.

The construction of a bird's wing is a glory of nature, the arrangement of the feathers producing an aerofoil that is strong, flexible and perfectly suited to its purpose. The usual arm bones – upper arm (humerus), forearm (radius and ulna), hand – extend from the shoulder. The main flight feathers – the primaries – come out of the hand. Ten of them, usually, attached to the hand with ligament and, to an extent, individually controllable. Next along, the secondaries are attached to the forearm, and the tertials – far less prominent – to the upper arm. The bases of all these feathers are protected by smaller feathers called coverts. Each flight feather overlaps the next, forming a stack that can be held tight to make a solid aerofoil, or fanned to allow the air to pass between the wing-tips. Some birds of prey have prominent 'fingers' at the tips, which can be held closed or open as necessary, giving them that little bit of extra control.

Another marginal gain is provided by the alula, or 'bastard wing' – a small group of feathers (usually three or four) on the leading edge, near the wrist joint. The alula is often deployed, along with other feather spreading, to help manoeuvrability at slow speeds, and especially when landing. It means the bird can descend more steeply and avoid 'wing stall'. This phenomenon – familiar to anyone who has ever held their hand out of the window of a car travelling at speed – occurs when the angle of attack of an aerofoil reaches a critical point, after which lift is drastically reduced. For birds, this is between approximately 15 and 35 degrees; for insects, it's greater – somewhere between 25 and 50 degrees. Bigger things generally stall at lower angles.

The size and placement of the alula puts it in the same category as the pterosaurs' pteroid or the dragonfly's pterostigmata – a small feature with an apparently disproportionate effect. The only birds without an alula are hummingbirds, but we've already seen how weird they are, and they seem to do perfectly well without it.

While a pigeon's wings offer firm surfaces for smooth air flow, they can also be flexed and distorted quite impressively. The scope to change the camber of a feathered wing is limited compared with the flexibility of bat and pterosaur membranes, but birds compensate for this with the ability to manipulate their primary flight feathers individually. Add to this their impressive range of rotation in the shoulder joint, through which they can change the angle of attack, and pigeons are well equipped to be strong, versatile flyers. This doesn't mean they're not averse to a bit of gliding. Their cousins, wood pigeons, are particularly adept at a specific display flight, flying at a sharp angle to crest an imaginary peak in the air, clapping their wings together at the top and then floating down with effortless grace. If it's half as much fun as it looks it would be reason enough to take up flying.

Pigeons also excel in the take-off. Being able to get yourself into the air at a moment's notice is an undeniably useful skill, and one common to many small-to-medium birds. In a lot of birds it happens too quickly for the human eye to perceive the exact order of events. But, much in the same way that the advent of chronophotography in the 1870s enabled us to see the order of the movement of horses' feet when galloping, slow-motion filming lets us break down the essentials of a bird's launch.

It starts in the legs. While the method – bipedal rather than quadrupedal launching – differs from that used by pterosaurs, the principle is the same: produce enough upward momentum with the push-off to enable a deep initial downstroke, without fear of the wings hitting the ground. And in pigeons, that downstroke is deep indeed, the wingtips meeting under the bird's body. Then the upstroke, again covering a wide angle. The wings meet at the top, clapping together audibly and producing the thrust-generating vortex of air we encountered in butterflies. The power generated by these wing strokes is substantial – enough not only to allow the birds to take off vertically, but also to accelerate from standstill to nearly 100 km/h in about two seconds.

For a bird the size of a pigeon, take-off is straightforward enough, even if it does require a short, intense burst of energy. But as body mass increases, it becomes more problematic, and the musculature required to propel a bird into flight needs to become larger. By the time you're the size of a swan, getting yourself in the air is a major operation. A mute swan (Britain's largest bird) needs a runway to take off. Visit a lake anywhere in Britain and you might be treated to the sight and sound of it – one of the great wonders of the natural world. The wet *smack-smack-smack* of their webbed feet pedalling on the water, accompanied by the *whu-whu-whu* of their wings pawing heavily at the air – all a reminder of the difficulty of flight.

But I digress. Blame the swans. They draw me in.

When it comes to landing, pigeons are as efficient as anything else at converting abundant kinetic energy to stasis. They might glide in, wings held in a sharp V and rocking gently on their axis as they ready themselves for touchdown. As they approach the ground, the feet go out and forward, preparing to cushion the landing. Meanwhile the tail fans wide, and the wings go into deceleration mode, flapping in reverse to create as much drag as possible. They execute the landing without the need for extra steps or overbalancing and are ready with their wings folded to get on with their regular pigeon business. To land on a ledge or perch, a pigeon will often approach at some speed from underneath and pull up at the last minute, with a drastic deceleration, placing themselves just above the landing spot so they can drop down gently and safely.

One of the pigeons outside the bakery, disturbed by a careless pedestrian, has just executed all this in front of me. The take-off, a swift go-around, back down for more scraps of discarded croissant. Casual as you like. I've watched it all the way, and now my eye is drawn back to the church tower, fifty yards away.

It's still there, a small, hunched shape in the cupola. As I watch, it takes off, dropping down from its perch and floating for a second before lifting itself up with deceptively lazy wingbeats and flying up and over the cemetery.

The peregrine flies high, flips gently backwards and round and down, and enters the 'stoop' – the dive that helps it attain those unimaginable speeds. Far down below, the pigeon flaps along without a care in the world, completely unaware that it is entering the last few seconds of its life. Pigeons, as we already know, are no mean flyers themselves, but even their speed and manoeuvrability might not be a match for a peregrine on form. The peregrine folds its wings into its body and accelerates, allowing gravity to do the bulk of the work. But it's not an uncontrolled plummet. With judicious adjustments of wing angles they can control the angle of attack to a certain extent. And, counterintuitively, it's thought that the faster the dive, the more effectively they're able to manoeuvre to intercept the erratic flight of their prey. But while the peregrine has the advantage in the dive, the pigeon's speed and agility give it the edge in level flight.

The speed of the stoop might cause damage to other birds, but peregrines have mitigating adaptations that allow them to attain these speeds without coming to harm. Like gannets they have nictitating membranes to protect the eyes from debris; their thick black malar stripe – the moustache-like lines descending from the eyes – reduces glare from the sun; and their nostrils have tiny bony structures that are thought to divert air flow and help them breathe more easily.

Shortly before it hits, the peregrine extends its talons. RIP pigeon. You were good, but sometimes good isn't quite enough.

The battle between peregrines and pigeons is a story of two birds with a long history of human domestication. Falconry has been part of our culture for at least 2,000 years, and while its practice as a hunting method has declined in many countries, it remains hugely popular in others, particularly in the Arab world. In Britain it was associated with the aristocracy from Norman times to the seventeenth century, but it became less popular from the eighteenth century as guns replaced birds of prey as the weapon of choice for hunters. It underwent a resurgence in the twentieth century, and now falconry is commonly practised, with the restriction that all birds must be bred in captivity, not taken from the wild. Our experience of falconry nowadays is mostly from public displays, where the birds are put through their paces, flying to the falconer's glove from designated perches, winning the admiration of the audience for their beauty and elegance, and just occasionally adding to the entertainment by decamping to a nearby tree and resisting all efforts to lure them back down.

The requirements for a hunter are fairly straightforward. Just do whatever it takes to catch your prey. This might mean being bigger, faster, cleverer, or more manoeuvrable than them, or a combination of any or all of the above. For peregrines it mostly boils down to one thing: speed. It's their USP. Fastest animal in the world and all that. Estimates of their diving speed vary from between 320 and 390 km/h, which, however you slice it, is on the nifty side. Their prey are predominantly birds – starlings, blackbirds, even the smaller gulls, and of course the abundant pigeons. But each to their own – different birds of prey use different techniques to outwit their meal.

For insight into a range of raptor techniques, take yourself, if you fancy it, to the Isle of Sheppey in December. I'll be the first to admit that, if you know this particular corner of the world just off the north Kent coast, this might seem a counterintuitive choice of winter day-trip destination. Marshy and bleak, Sheppey lies at the point where the end of the Thames meets the beginning of the North Sea. In December, daylight is short, and the weather is likely to be – not to put too fine a point on it – vile. Low clouds, stiff winds, cold, grey and damp. The kind of weather you send to your enemies.

You might, while you're there, make a pilgrimage to Halfway Cemetery, there to pay homage to our friend Uwe Johnson – subject of the museum in Klütz – who, to the bewilderment of all who knew him, chose to spend most of the last decade of his short life on Sheppey. 'It is not pretty,' he wrote to a friend. 'But is that why we've come?'

On balance, probably not. On Sheppey, desolation is your friend. But so are the marsh harriers, barn owls, short-eared owls, sparrowhawks, peregrines, merlins and kestrels, and they're the reason you've come.

You're heading for one particular spot, on the Harty Ferry Road. Long known as a fine place to watch birds of prey, this is now officially designated the 'raptor viewpoint'. You arrive, maybe an hour before dusk – in mid-December that's basically just after breakfast – and you wait.

The marsh harriers might be first. Large birds, long and broad of wing, there's a slow menace to their languid flight. At the wing-tips, the feathers are spread into recognisable fingers – closed for smooth flight, open to slow down or manoeuvre. They come here in numbers – ten, fifteen maybe, twenty if you're lucky – floating low over the reed beds, occasionally banking off to one side and going around and up before dropping again to resume the search. When in full hunting mode, they work their way methodically over the wide expanse of reeds – quartering, to use the proper term – scanning for small mammals. Sometimes they will glide or soar, the wings held up in a shallow V. Many migrate to Africa, but here they stick the winter out. (There's loads of prey on Sheppey.) In spring, their spectacular courtship display catches the eye, both parties executing deep dives and rolls and gyrations, the male staging mock attacks on the female. Sometimes he drops food for her to catch, no doubt confident in her ability to pull off that particular feat of mid-air dexterity. 'Sky dancing', they call it. It would cost you twenty quid for the cheap seats in the West End – but it's free on Sheppey.

All that is to come. But for now survival is the main aim.

The drop is sudden. One less marsh harrier in the air. The air's loss is the reed bed's gain. Something is dying. A rabbit maybe, or another small mammal. Poor thing.

Tear your attention from the marsh harriers, and in the gathering gloom, off to the left, you see a ghost. Barn owl. A proper magic bird. It's not just the pale cream and honey of its plumage, nor the heart-shaped face, delightful though both those things are. It's the silence.

It's easy to forget that birds' wings make sound. From the frenzied *whirr* of a robin fleeing a cat to the *whu-whu-whu* of those mute swans, the displacement of air at speed is audible. Owls are an exception. For that they can thank their feathers. Of course they can. We've already established that feathers are magic.

The feathers of a barn owl are soft and light, the feel of their plumage often compared with velvet. And their secret weapon is the construction of the most prominent primary feathers. The leading edge of these feathers – the one that first meets the air – has a serrated edge, some of the barbs longer than others. Although the exact mechanism isn't known, the effect of this comb-like fringe is to change the air flow around the wing and deaden sound. The trailing edge also has a discernible fringe. So effective is this adaptation that, in tests, microphones were barely able to measure the level of sound emanating from a flapping barn owl. Elsewhere on the owl's body, the contour feathers are also built to ensure air flow over the bird is as smooth as possible. They have a light body and large wings, so they can fly extremely slowly without dropping out of the air and can even muster up a light hover with the support of a little updraft.

Add acute eyesight and hearing to their stealth capabilities, and you have a highly efficient hunter. One ear is higher than the other, the better to detect and pinpoint any sounds made by its prey, particularly those in the high frequency range – creep softly, little vole. And those big, forward-facing eyes are great not just for depth perception but also for detecting movement in low light, conditions for which Sheppey in December is noted. The silence of their flight is a double bonus: the hapless vole can't hear the owl, and the owl can hear the hapless vole.

There is a downside to their highly modified plumage. They don't like rain. Those delicate feathers are easily waterlogged, so in the wet they become both less efficient and noisier. But what they can handle is wind. Owls have excellent head stabilisation. The body can be buffeted, but the head remains absolutely motionless. The body absorbs the gust automatically – a mechanical response, allowing the nervous system to catch up and adjust accordingly.

Also masters of the still head are kestrels, those motorway familiars, hovering at height above the verge while keeping their eye fixed on the path of a vole or mouse before dropping down to execute the kill. And in flight they're fleet, nimble, darting on pointed wing in pursuit of small birds. They'll catch them, too, likely as not, although for aerial agility they yield to their cousin, the hobby. So manoeuvrable are they that catching dragonflies is routine – hunters can also be hunted – and they are possibly the only bird capable of taking a swift in mid-air, a task beyond even our speedy peregrine, which flies strongly and nimbly enough, but will typically fly after its prey only in the event of a failed dive.

Nearly all raptors incorporate birds into their diet, and some eat almost nothing but. With all these potential killers about, life for a smaller bird is fraught with peril. How can they stay safe amid such danger?

They have warning systems in place, of course, their alarm calls alerting anything in the vicinity to the presence of a predator. Sometimes, though, it's best not to draw attention to yourself – there's nothing quite like the abrupt silence of chattering garden birds when a sparrowhawk turns up.

If they can't avoid detection, the simplest way to avoid danger is to fly, fly far, fly fast. And this is often effective, especially if the smaller bird can find a hiding place out of their pursuer's reach – in contrast to the extraordinary hunting efficiency of dragonflies, most bird-of-prey hunting attempts end in failure. But all a peregrine needs to do is catch one pigeon and that will do it for the day. As already advertised, there are a lot of pigeons in cities.

Another effective strategy against predation is to gather in large numbers. Presented with a swirling mob of moving targets, the predator doesn't know what to attack. They might strike lucky and take out one of the flock, but that's a price the flock's prepared to pay. Safety in numbers – except for the one sacrificial individual.

Loose flocks of urban pigeons – sweeping round at roof level, changing direction unpredictably, apparently carried by whim and exuberance – catch the eye. But if you're after astonishing formations of endlessly morphing, shapeshifting billows of bird, starlings are the masters.

The gathering of thousands of starlings towards dusk on winter evenings provides a spectacle of aerobatic brilliance for anyone who cares to be in the right place and has the inclination to look upwards. The bird cloud seems to take on a life of its own as it shifts and surges, sometimes almost seeming to breathe in and out like a giant lung. It's a breathtaking spectacle, but an aural treat too – the fast breath of movement produced by the whirring of thousands of wings.

If it's windy or cold, the starlings still gather, but tend to drop down into the roost without fuss or hesitation – the warmth created by the density of birds is important for their survival. But when they do perform, it's impossible to tear your eyes away. It sometimes feels as if you're watching a computer-generated waveform, each element programmed to obey certain rules. And in a way, that is what appears to be happening. Each bird is responding to the seven or eight birds around it – if one of them moves, it follows. This principle is simple enough, but how they do it without crashing into each other seems extraordinary to us. Part of that can be explained by the birds' speed of reaction, which makes a mockery of ours. So what seems instantaneous to us happens in 'normal' time for them. Imagine something with reaction times many factors slower than ours sitting in the roof at Waterloo station and wondering how the commuters manage to avoid each other – that's us watching starlings, although of course starlings don't murmurate with their heads in their phones.

It's one thing knowing how each individual bird behaves, but when you extend it to hundreds, thousands of birds, the mystery deepens. Never mind that we still don't know which bird, if any, makes the first move. There is no discernible beginning, and no discernible end either. But how the mass of birds – a complex, dynamic system – retains its shape, always active and always potentially on the brink of collapse, is still poorly understood.

The disruption to the group caused by a predator attack only adds to the spectacle. It billows, splits and undulates as the raptor dives into the throng and individual birds try to escape it, with the usual knock-on effect. They're moving in such numbers and so unpredictably that the predator's strategy is to aim for a point in the mass rather than an individual bird. They're most likely to catch anything that is moving on the same plane as them. The remarkable thing about this is not just the kinetic shape of the group as the disruptor ploughs in, but how quickly it reforms. And it's a happy coincidence that they form uncanny three-dimensional patterns that provide us with enough awe and fascination to keep us warm on a frigid winter's night.

Starlings are matched in unanimity of purpose by several wader species, such as knot or dunlin, which put on similarly mesmerising winter displays over expanses of mudflat. They rise as one, a billowing sheet of birds, flying close to the ground then waving upwards, the appearance of the flock changing from dark to glistening white as they bank and roll, their undersides picked out by the low winter sunlight.

While such displays might stem from the collective desire to thwart potential danger, birds sometimes use more direct action to see off an interloper. This behaviour is often seen even with small numbers of birds – jackdaws chasing away one of our local peregrines is the one I see most often over my south London home – but sometimes the mismatch of size is even more stark. The fearlessness of the dozen or so great tits (14 cm) I saw mobbing a buzzard (57 cm, and in possession of talons not much shorter than the entire body length of its attackers) over the Kent countryside one summer afternoon cannot be overstated, and can serve as inspiration to anyone supporting the plucky group of outsiders taking on the big bad guy.

In these encounters, the smaller birds seldom make physical contact with the larger, rather choosing to make themselves enough of a pain in the arse that the interloper loses the will to fight. They harry, they swoop, they dive-bomb, and the result, more often than not, is that the larger bird moves on without carrying out its master plan.

Sometimes a flock is formed for another kind of survival: feeding. Many smaller birds use flocking as a foraging strategy, especially in winter, when food might be scarce. On these occasions it's common to find flocks of mixed species, a counterintuitive finding for anyone who has ever come across the generalised saying 'birds of a feather flock together'. As well as protecting from predators in the usual way, it's suggested that flocks of all kinds might also have a social purpose – the 'information-centre hypothesis' suggesting that they use it as an opportunity to share information about the best foraging sites. This behaviour benefits the whole group – except when there's not enough food to go round, in which case it's the weaker members of the group who miss out.

'Twas ever thus. Life is tough.

From peregrine to pigeon, hummingbird to goose, birds represent the gamut of flight styles and capabilities. They hover, glide, dive and soar – some of them just flap and flap until they've got to where they need to go. They use their ability for the briefest of journeys – a flit from perch to tree – and the longest – across continents and from pole to pole. Flight enables them to eat or to avoid being eaten. Sometimes it seems they do it just for fun. For many, birds are the embodiment of flight, the answer to the question 'Can you name an animal that flies?'

Flight gave them the world, and they accepted readily.

With the pterosaurs out of the way, birds became the only flying vertebrates, monopolising the 'medium to large' size bracket and exploiting nearly every available niche.

Nearly.

But they left one niche largely unexplored, and before long another kind of animal evolved to take advantage of that gap, bringing flight to the world of mammals.

14

THE BAT

Mid-July. Evening. The air has the kind of unthreatening warmth that invites dawdling. There is no reason to move indoors, every reason to stay outside and allow peace to drape itself over you as night falls. The low hum of insect activity in the garden has provided a pleasant backdrop to our meal, and now, bit by bit, as the air cools and the light fades, it slows and quiets, and we're left with a screaming party of swifts, terrorising the neighbourhood with their banshee wails. But soon enough they realise it's time for bed, and now there's just the low background hum of London, and the stillness of gathering night. I glimpse a shape flitting around at treeline level. It is doing it in a way that tells me it is not a bird.

A bat. Most likely a pipistrelle species (common or soprano), the default small bats of Europe and the two most abundant in Britain. Get a common or soprano pipistrelle up close and it reveals itself as an endearing, furry creature – no more than 8 or 10 cm long – with a short muzzle, prominent, rounded ears, and small black eyes. But for most people it's encountered like this, as a flickering presence on a warm summer evening.

I trace its progress around the treetops as well as I can, my eyes adjusting to the dark well enough to follow it even when it isn't clearly silhouetted against the blue-purple of the darkening sky, or at least to guess where it might appear next and catch up with it if my guess is slightly out. Its speed and the unpredictability of its path make this a tricky endeavour, and before long it loses itself in the darkness. I head indoors, pleased with the encounter, but, as always with bats, wanting slightly more.

There is an air of mystery surrounding bats. This is entirely understandable. Such is the fate of anything nocturnal – lives lived largely out of our consciousness, our worlds coinciding with theirs only rarely and fleetingly.

They're worth knowing though. After all, they are, of all the things that fly, our closest relatives, and the only mammals to have evolved powered flight. That in itself is worthy of note. If we could see them more clearly, their flying exploits visible to all during the hours of daylight, they might just be as popular as birds.

It might be different, too, if we had better hearing. The high-pitched sounds produced by bats – from about 11 kHz to 212 kHz[1] – overlap only to a small extent with the upper reaches of our hearing range (up to about 20 kHz, but shrinking as we get older). A child, or a particularly keen-eared adult, might hear the contact calls of a common or soprano pipistrelle at 45–55 kHz, but for the most part their communications go undetected, except to those with specialist equipment. If we could hear them, how it might alter our perception of them.

And the extraordinary workings of echolocation – the use of rebounding sound waves to locate objects – open up a world of endless fascination. The notion that bats used some other sense than sight to manoeuvre in darkness was first proposed by Italian priest and biologist Lazzaro Spallanzani in 1793. He noticed that while a barn owl bumped into obstacles when flying in complete darkness, a bat didn't. The experiments he and other scientists used to explore this hypothesis – including the removal of eyes and piercing of eardrums – are blood-curdling to modern sensibilities, and while he remained in the dark about the exact workings, he discovered enough to be convinced that hearing was at the heart of it, despite the mockery of his contemporaries (Baron Cuvier, of 'ptéro-dactyle' fame, reportedly remarked, 'Mr. Spallanzani, if bats see with their ears, do they hear with their eyes?')

It wasn't until the twentieth century that an explanation was found, first with Hamilton Hartridge's proposal of ultrasound in 1920, and then with experiments by Donald Griffin eighteen years later. Using a device developed by George Pierce – a 'sonic receiver', precursor to modern bat detectors, which converted high-frequency sounds so that they were audible to humans – Griffin and Robert Galambos discovered that bats were producing sounds in flight, and when deprived of the ability to produce or hear them, they lost their seemingly miraculous ability to avoid obstacles. It was Griffin who coined the term 'echolocation', and his continued experiments led to further discoveries about the remarkable phenomenon central to the lives of these extraordinary creatures.

Bats were the last of the world's flyers to develop the ability. It's no coincidence that it emerged relatively shortly after the K–Pg extinction. The planet had seen periods in which various animal life forms had thrived. Now was the time for mammals to come into their own. The survivors of the extinction picked up the pieces and made their way in an unfamiliar world. The absence of huge swathes of animal life meant that niches abounded, just waiting to be filled. So they filled them.

The birds did well, as we know. Snakes, turtles and crocodiles all made it through more or less intact. Mammals, too, lived to fight another day, diversifying and radiating rapidly. Of the many mammal families that spread around the world, only one group thought to take to the air with powered flight. How they did this remains a mystery. There is no *Archaeopteryx* for bats, no transitional fossil to help make sense of the progression from terrestrial to aerial. It's thought there were protobat species in the late Cretaceous, just before the extinction, but the first true bat fossils date from about 52 million years ago. Unlike with the Hexapoda gap in the fossil record of the insects all those millions of years earlier, the fossil gap between those protobats and the true bats is a mere 15 million years, but that's enough for questions to abound. How, why, when?

Delicate of bone and thin of membrane, bats fossilise especially poorly. There are far fewer complete skeletons than there are oddments of tooth and bone, so even though the bat fossil record is relatively recent, the jigsaw puzzle is in many ways more difficult to put together. But we do know that bats are monophyletic – that is, if you were to go up to the putative Tree of Life and snip off the bat branch, you would have bats and nothing but bats, and no bats would be left on the tree. The exact nature of the common ancestor of all bats can only be presumed: small, probably; nocturnal, probably; tree-dwelling, probably.

The earliest undeniable bat fossil is called *Onychonycteris finneyi* ('clawed bat named after Finney').* It's known from two complete (albeit rather rumpled) skeletons from the Green River Formation in Wyoming,† laid down 52.5 million years ago. Based on a reconstruction of its aerofoil, *Onychonycteris* seems to have been a weak flyer, having to flap hard and fast to generate its undulating, fluttering flight, and interspersing these efforts with periods of gliding.[2] Unlike modern bats, which retain two claws per hand at most, it was fully clawed, indicating that it was probably a strong climber. And from its skull anatomy – specifically the small size of its cochlea, the part of the inner ear involved with hearing – it can be extrapolated that it was in all probability not adapted to use echolocation. Contrast this with another bat fossil from the same formation. *Icaronycteris index* was smaller (about 25 g to *Onychonycteris*' 40 g), and its anatomy and proportions more closely resembled modern bats. With a comparatively large cochlea, it's reckoned to have been capable of echolocation as well as powered flight.[3] Elsewhere, Germany's Messel Pit, from around the same time, boasts a wealth of bat fossils (more than a thousand specimens from at least eight species), some of which are preserved in such detail that their stomach contents are visible. These details can be helpful in determining an animal's lifestyle

* Bonnie Finney, who discovered the fossil.

† Also, you may remember, the venue of the hummingbird ancestor *Eocypselus rowei*.

and feeding habits – the variety found in the Messel Pit is a sign that bat diversification was already well under way.

The question of how flight developed in bats throws up very similar arguments to those concerning the origins of flight in birds. Ground up or trees down? As *Onychonycteris* and *Icaronycteris* were climbing tree-dwellers fond of a glide, this would seem to favour the trees-down hypothesis. And their contrasting echolocation skills add fuel to the debate about which came first: echolocation or flight?

As so often, there are conflicting hypotheses. If you subscribe to the idea that echolocation developed first, the progression goes something like this. An animal sits on a perch, reaching out for passing prey with long limbs. It detects its prey with keen eyesight and a strong sense of smell, and communicates with ultrasound. Over time, the animal's descendants develop longer limbs, and webbing grows between their fingers to make it easier to scoop up passing insects. The webbing becomes a full membrane supported by the fingers. As their reach gets bigger, their need to locate their prey becomes more important, so the ultrasound becomes more sophisticated, turning into a form of echolocation. Then the animal starts to jump from its perch to catch more prey, and so on. Although this hypothesis is backed up by the existence of echolocation without flight in some shrews, it depends largely on the idea that protobats commonly used reach-hunting, a behaviour not observed in modern bats in the wild.

The 'flight first' hypothesis seems to have just as much going for it. It charts a relatively simple progression for a tree-dwelling animal: jump, glide, fly. There's support for it in the observation of gliding mammals such as colugos.

Colugos are also known as 'flying lemurs'. This – following the example of the horse chestnut, which is neither a horse nor a chestnut – is one of those misnomers that cause widespread perturbation. While they are closely related to lemurs, they take the winsome, big-eyed appearance of those Madagascan primates and raise it to the power of five. Colugos deserve recognition in their own right, not least for their aerial abilities, in which department they definitely have it over their cousins. What big eyes you have. All the better for seeing in the dark, obviously. And what a big patagium you have. All the better for gliding from tree to tree.

See a colugo clinging to a tree and you might feel sorry for it as you watch it climb with laborious and measured drag-jumps, rather in the manner of a weakling child climbing a rope.* But when it needs to move from tree to tree, it comes into its own, flinging wide all four limbs and gliding gracefully, almost with a sense of abandon, to the next tree. The colugo maximises the potential of the patagium, the tip of each extremity – fingers, toes, tail – joined by a vast sheet of membrane to make an aerofoil that makes it the most effective of all the gliding specialists. When not in use, the folded membranes do give the animal a slightly baggy appearance, calling to mind Edward Newman's possum-like representation of pterodactyls in the nineteenth century, but as it's rarely in a hurry to get anywhere unless it's gliding, it's barely a hindrance.

A colugo in mid-glide is a reminder of the potential, given the right configuration, of thin, stretched skin as an effective aerofoil. While they have so far reached only the second stage of the jump-glide-fly progression proposed for the evolution of flight in bats,

* That was me, in case you hadn't guessed.

they could easily be representative of a protobat that has worked out gliding but is still to crack the flapping bit. Despite the similarities, however, neither they nor flying squirrels are bats' closest relatives – that honour, slightly surprisingly, seems to go to the group including horses, pangolins, whales, cats and dogs.[4]

There's more support for the jump-glide-fly hypothesis in the connection between echolocation and flapping. Bats using laryngeal echolocation synchronise the sounds they produce with both their outbreath and the upstroke of their wingbeat. This effectively ensures that the sound production costs them no energy. To produce a pulse of ultrasound while sitting still uses a lot of energy – tie it up with something you're already doing, like flapping, and it comes free of charge. Ergo, flight must have developed first.

Both these hypotheses have been subject to criticism, however, so fans of the middle ground will be delighted to learn that, as in the debates over the origin of flight in insects and birds, there is a third hypothesis – in this case it proposes that the two evolved together, the jumping-gliding-flying progression moving in step with the advance of the bats' echolocation technology. As they became stronger flappers, sound production became easier. A win all round. And as they became larger, their wingbeat frequency became slower, and along with that the frequency of echolocation calls decreased and became less useful. So this proposal is also at least a partial explanation for the general lack of echolocation in larger bats.

However it came about, the evolution of flight no doubt played an important part in the success of the bats. Once they got going, they radiated far and wide. Gondwana was well broken up by then, but the resulting landmasses were still close together, so dispersal for flying animals was no big deal. Now we have bats on every continent except Antarctica, with the highest concentration in tropical and subtropical regions – Colombia has the highest diversity, with about 180 species. Bats live in varied habitats – trees, caves, burrows, sometimes in your porch – and eat a wide range of foods, from insects to moss, nectar to blood. They're the second most numerous mammals, after rodents, with more than 1,400 species worldwide. They range in size from the tiny (Kitti's hog-nosed bat, also known as the bumblebee bat, weighing in at about 2 g) to the hefty (several species of the fruit bats known as 'flying foxes' nudge the 1.5-kg mark). Most fall into the lower range, below about 50 g.

The classification of bats, like the classification of most things, has not been without its difficulties and confusions. For a long time the divisions seemed clear enough. There were megabats (Megachiroptera) and microbats (Microchiroptera). The megabats included the flying foxes, feeding mostly on fruit (hence the general name 'fruit bat'), and orienting themselves in the world using sight and smell rather than echolocation. There was one genus that used a basic form of echolocation with tongue clicks, but what we regard as 'true' echolocation was the preserve of the microbats. These were generally (although not always) smaller than megabats, and lived mostly on insects. This was accepted as the status quo for years.

Then things, as they tend to, changed. New discoveries in the 1980s and then in the early 2000s showed that some of the 'microbats' were more closely related to the flying foxes than previously thought, and so there was a general reshuffle and two new orders were created. Orders, I should note, with much longer names, which, having told you what they are, I will arbitrarily and unscientifically shorten, for the sake of the general mental health of the nation. The Yinpterochiropterans ('Yin') were the Bats Formerly Known as Fruit Bats plus a few others, including horseshoe bats and mouse-tailed bats; the Yangochiropterans ('Yang') number nearly a thousand species, and generally align themselves with the group formerly known as the microbats.

A very few bats – the rousette bats – use the tongue-clicking method of echolocation mentioned earlier. This is a technique that a human could probably emulate with a bit of practice. It involves clicking your tongue, listening to the sound returning from whatever it's bounced off, and building a rudimentary picture of the world from the information you receive. By moving and comparing, you might get good enough at it not to run into the wall. This type of echolocation is limited in its scope; it is really most effective if your prey – probably an insect – is sitting nice and still on a leaf fairly nearby.

The majority of bats do it differently. Not only is the means of sound production different – they produce the sounds from their larynx, which is specially adapted to produce the high-frequency sounds they need – but they have a much wider range. The lowest frequency mentioned above, 11 kHz, is used by a moth-hunting bat, *Euderma maculatum*,[5] its motivation being to evade detection by the moths. And while the frequency produced at the top of the range, by Percival's trident bat,[6] is a handy demonstration of the narrowness of our own perception of the world, most echolocating bats settle for a range between 20 and 60 kHz, the precise span varying according to species.

There is enormous variation in the exact kinds of echolocation used by bats, but a typical cycle might start with a search call – comparatively long, low in frequency and with a low rate of repetition. Once an object is detected, the bat reduces the intensity and duration of the call while increasing the repetition rate and frequency. In the final phase, as they home in on their prey, there is a sharp increase in the repetition rate to produce what is known as the 'terminal' or 'feeding' buzz. Even in this stage, most bats emit the call and wait for the echo before emitting the next one. Some, though – including the greater and lesser horseshoe bats found in Britain – are able to broadcast and receive at the same time, differentiating the call from the echo by frequency. The echo from moving prey returns to the bat at a lower frequency than the outgoing call, thanks to the phenomenon known as Doppler shift (the most familiar manifestation of this can be observed by listening to the change in pitch of a passing ambulance siren).[7]

The high frequency of bat calls goes hand in glove with short wavelength, which reduces the chances of prey slipping through the acoustic net. The problem with high-frequency sounds, though, is that they don't travel far. Bats compensate for this with calls of extremely high intensity – up to 140 decibels, which is considerably louder than your fire alarm, and enough to cause pain if we were able to hear it. This in turn yields another problem: how to do this without deafening themselves. And here's the clever bit: they automatically engage three bones in the inner ear, switching them on and off in synchronicity with the shouting.

It's not all shouting though. Some bats – known as 'whispering bats' – echolocate quietly, flying close to trees and shrubs and picking insects off the leaves. And some nectarivorous bats also use echolocation to find nectar. They're helped in this endeavour by the plants. *Marcgravia evenia,* a vine endemic to Cuba, is one example, its concave leaves reflecting the bat's calls and guiding them towards it.[8] This arrangement is of mutual benefit, the bat emerging from the encounter – like hummingbirds in similar circumstances – covered with pollen.

While echolocation is a powerful tool, bats don't have it all their own way. Some insects can hear them coming, and are able to take evasive action. Others, such as tiger moths, respond with their own sonic signals, emitting a series of clicks which either warn the incoming predator that they're not good to eat – the aural equivalent of the gaudy aposematic colouring widely used by insects – or jam the incoming bat calls.[9] Despite these challenges, echolocation has enabled bats to thrive in an otherwise unoccupied niche, taking advantage of the abundance of nocturnal insects without competition from other predators. It allows them to build an accurate acoustic picture of the world around them, without the need for sight (although most of them see perfectly well). Now all they need to do is manoeuvre their way around it.

And when it comes to flight, they have their own special way of doing things.

For insectivorous bats, manoeuvrability is the gold standard. Being able to locate an insect in the dark with your special sound-bouncing kit is good; being able then to home in on it and catch it is even better. With insects displaying their own brand of virtuoso flying, bats need to be able to outmanoeuvre them if they're going to make their way in the world. Key to this is what they can do with their wings. Compare them with birds and pterosaurs.

Birds have that strongly keeled sternum for the anchoring of flight muscles, a feature that is smaller in both bats and pterosaurs. Bats make up for this with a more complex system of flight muscles. The pectoral muscles, while not as chunky as those of birds, nevertheless bear the brunt of the downstroke, but they are supported by a network of arm muscles that help bats control and manipulate their wings with great flexibility.[10]

This is enhanced by the bone structure of the wings. The bones in a bird's wing go only so far, feathers taking over where the hand leaves off. Pterosaurs had that ridiculously elongated fourth finger. Bats, however – with, incidentally, exactly the same hand-bone structure as humans – have their fingers well embedded into the membrane, spreading through it right into the tip. So they already potentially have more control over the aerofoil. Factor in a few more elements: they have intricate shoulder joints, which allow for more rotation; their wrists and elbows are foldable; the finger bones are flexible; and the wing membrane is not just an inert flappy sheet of skin but shot through with tiny muscles. Combine that with the bones, and this enables them to manipulate their wings with deftness and precision. The flexibility does mean that the wings are more liable to flutter than the relatively rigid surface of birds' wings, but this doesn't seem to have an adverse overall effect on lift production, and the complex vortices swirling across the upper surface of the wing might play a part in the generation of thrust.[11] Another potential disadvantage of the membranous wing might be its vulnerability to damage, but the membrane has a surprising tolerance for wear and tear, and heals quickly. Meanwhile, the bat can go about its usual business, not apparently inconvenienced by the wound, unless (as is sadly often the case) it was caused by a cat. Much-loved family pets as they are, cats are responsible for most bat predation in the UK – the wounds they inflict can lead to infection, a major obstacle to healing.

Another factor in a bat's aerodynamic performance is fur, the extent and nature of which varies greatly between species. The hairless bat (*Cheiromeles torquatus*) of southeast Asia – also known, charmingly, as the naked bulldog bat – is, as you might deduce from its name, entirely smooth-skinned. At the other end of the spectrum are some of the roundleaf* bat species in the genus *Hipposideros,* which have rough, dense fur. The potential advantages of hirsuteness include insulation and camouflage, benefits that outweigh the slight loss in aerodynamic efficiency. There is also a broad correlation between length of fur and wing shape – a bat with short fur is more likely to have long, narrow wings. The wings themselves lack fur, but are thinly covered with sensory hairs that play a crucial role in flight control and stability[12] – their removal results in reduced manoeuvrability.[13]

Regardless of those marginal factors, it's the flexibility of bats' wings that accounts for much of their aerial prowess. In particular, the ability to fold their wings on the upstroke and slide them up close to the body is a huge factor in drag limitation. And while some bats are capable of fast flight – the fastest recorded was 160 km/h by a Brazilian free-tailed bat, although as always with such measurements there are caveats about their accuracy – some can also flap very slowly and still generate significant lift with a

* The word 'roundleaf' refers to the shape of the flesh around their nose. In some species, such as the greater horseshoe bat, this distinctive appendage serves the purpose of directing their echolocating calls.

leading-edge vortex. Hunting over water, little brown myotis bats were measured at speeds as low as 8 km/h.[14] In such circumstances, the bat needs superb control of its wing strokes – clipping the water is not advised – so will reduce the depth of the stroke while increasing its speed.

Contact with water isn't such a problem for the few bat species that specialise in hunting fish. The aptly named fish-eating bat (*Myotis vivesi*), resident of the Gulf of California, is an obvious example. They have elongated claws with which they can snag any fish or crustaceans carelessly hanging around near the surface. Their long wings, high in aspect ratio and low in wing loading, give them the perfect blend of speed and agility.[15]

Another part of the secret to a bat's mobility lies in the weight of their wings. Again compared with birds, the wings of a bat comprise much more of their body weight. This might seem like a disadvantage – surely you want wings to be light? But that weight gives them the benefit of inertia, which they can harness to manoeuvre themselves more deftly, as well as making them more resistant to disturbance from gusts of wind.

This manoeuvrability manifests itself when a bat comes in to land. Whether executing the two-point or four-point touchdown, agility is key. The most spectacular landings involve a somersault on approach. They slow down, do a back flip, stretch out their feet and grab, all within the last couple of wingbeats. This expertise comes in handy whether they're negotiating their way into a tight crevice – as with all UK species of bat apart from the two types of horseshoe bat – or manoeuvring themselves into the trademark upside-down position, with wings folded to insulate the body and protect the wing membrane. That roosting position is enabled by a special tendon (not dissimilar to the one in a perching bird's leg) which lets the feet lock shut for energy-free suspension.

From here take-off is simply a case of letting go and flapping – the higher they roost, the more speed they can build up in the descent. Bats can't run, so launching from the ground, while not impossible, is not their preferred method. The common vampire bat, which feeds on or near the ground, is an exception. To facilitate the launch, they jettison as much weight as possible by urinating, then crouch down, extend their wings and use their unusually long thumbs to fling themselves into the air, as if executing an especially powerful downstroke. This is not dissimilar to the proposed launch mechanism used by pterosaurs, but bats' legs, puny in comparison to the powerful launchers of pterosaurs and many birds, are of little use in the process. Slender and delicate, they're good for hanging from, and not much else.

Once in the air, a bat uses a lot of energy. Flight, as so often advertised in these pages, doesn't come cheap. Chasing after insects – with all the ducking and diving and fine motor control involved – is extremely energy-intensive, as is hovering to drink nectar. A flying bat consumes approximately twelve times more energy than when not in flight.[16] It helps, then, that bats have more efficient metabolisms than most mammals. An insectivorous bat might eat its own body weight in insects every day – perhaps more[17] – so it makes sense for them to feed where food is abundant. A densely packed swarm of insects represents an easy opportunity to consume a good percentage of your daily allowance. Some consume insects on the wing, but others will catch them and then take them to a perch where they can tuck in at their leisure. The advantage of this approach is that it enables them to save energy by catching larger prey.

As well as having to counter the disadvantage of a mammalian metabolism, bats are also encumbered with a mammalian respiratory system, which they have developed to cope with the demands of flight. It's not of the order of the one-way air-sac system evolved by birds, but with enlarged lungs and hearts, they're able to work at a high and sustained metabolic rate.

The limitations of the mammalian respiratory system might be one of the reasons that bats haven't grown any larger, as well as an explanation for the reluctance of other mammals to join them in the air. Flying gets harder as you get bigger, because of a simple equation of scale. Double something's length, and the surface area becomes four times as much while the mass goes up by eight times. Birds can (and pterosaurs could) cope with those extra demands thanks to their ultra-efficient respiratory systems. Bats don't have that advantage, hence (and this might be an enormous relief to anyone who finds them creepy) the upper limit on weight, as demonstrated by the flying foxes. One and a half kilograms isn't much compared to some of its fellow mammals, but then they don't often flap around just above head height making unpredictable changes in direction the way flying foxes do.

Within the limitations of mammalian anatomy, bats have seized the opportunity, dodging competition from similarly agile birds by becoming predominantly nocturnal and making the night skies their own. There are a few diurnal bats (but, in case you were wondering, no flightless ones), but the best opportunity to observe them often comes at the fringes of the day, those moments of incipient daylight or gathering darkness, filled with shadows, when you can't quite be sure what's what.

Just occasionally, if you're lucky, a bat will find its way into the house, and you'll become aware of its presence, its wings making a faint whirr as it does endless laps around the central light fitting, unerring in its path and apparently disinclined to deviate from it. Round and round and round and round, with boundless energy, both powerful and insubstantial, possessed of a strange vulnerability. You might be able to make out the shape of its wings, the thinness of the wing membranes. You'll watch for five, ten, fifteen minutes, transfixed, wanting to be able to hear those ultrasonic sounds, just for a sign of proof that it actually exists, because while you know the details of what it's doing, it still defies belief that an animal can exist in that realm, and you still don't quite trust, or, truth be told, understand it.

Then, for no apparent reason, it breaks free from the endless circling, finds the window, and is gone, leaving you both slightly bereft and subtly enriched.

AFTERWORD

Life is tough. Wherever you are in the food chain, whether predator or prey, the forces of nature conspire to thwart you. Either something is always trying to eat you, or the things you want to eat don't want you to. Survival is the only game in town, whether you spend your time in the water, on land, or in the air.

Those insects that first took to the air were, did they but know it, really onto something. It opened up new possibilities, vistas of opportunity. Escape, freedom, safety. Not to mention the views.

And then others discovered it, finding ever more inventive and outlandish ways to do it, and the novelty wore off. But still. There is something cussed and defiant about becoming a flyer. On our planet, gravity is powerful. It looks at us and says, 'I'm going to make sure you stay down there, where you belong.' Flyers aren't having any of it. 'Fuck you, gravity,' they say. 'I'll do what I please.'

From the moment when that first insect flapped its drawing-board wings and found itself, however fleetingly, going up instead of down, the air has been open to anything with the will to work for its aerial reward. It's far from easy. Defying the forces of nature often is. More often than not there are compromises to be made in the pursuit of aerial facility. It's not for all. Some are destined never to make it. They're too heavy, too cumbersome, too leaden-footed. Others manage it but decide after a while that they don't need it. Yet more engage in it only fleetingly, unwillingly, or half-heartedly.

There are others that, for all we know, are on their way to developing it. Flight is a continuum, from the very weakest to the very strongest, and while gliding doesn't conform to the definition 'powered flight', it does conform to the definitions 'moving through the air without plummeting to an untimely death' and 'staying off the ground for longer than you might reasonably expect, given what you know about the forces of gravity'. Flying squirrels, flying lizards, flying frogs, flying fish, flying snakes – all, to the purist, misnamed, but all using their bodies to exploit aerodynamics, and exerting enough control over their descent to avoid crashing into things and manoeuvre themselves to a safe landing place. Each has their own idiosyncratic style, from the stiffened Chitty-Chitty-Bang-Bang wing membranes of the Draco lizards to the flattened wriggling S-shape adopted by the Chrysopelea snakes. Their landings might not always be the height of elegance, but for them gliding is an entirely worthwhile adaptation – and who knows how they might develop it in the future? For all we know, the gliders of today are the flyers of tomorrow.

For those animals that did develop it, flight proved the key to survival. All the things in the air today owe their existence at least in part to that strange, apparently unnatural ability.

It has enabled insects to become the most varied and numerous life forms on the planet. The pterosaurs joined them in the air, and explored the possibilities of vertebrate flight, taking the mechanics of it to the very limits. Damn that asteroid. For birds – the only living dinosaurs – it was everything. And for bats it has been the key to domination of the night skies and exploitation of a niche unexplored by other life forms.

Flight, in short, is quite the thing.

'We have to continually be jumping off cliffs and developing our wings on the way down.' Kurt Vonnegut, there. He knew a thing or two. Today, there are millions of animals doing it, each with its own idiosyncrasies of technique – from the feeble and unwilling flight of craneflies and tinamous to the apparently effortless mastery of dragonflies and hummingbirds – and each one descended from something that jumped off a cliff and developed its wings on the way down. And they have one thing in common.

They fly.

They. Fly.

And no matter how inefficiently they do it, no matter how clumsy they look, no matter what the personal cost, flying is, above all, cool. Far cooler than running or walking or swimming or slithering. And wouldn't we just love to join them up there?

One of them went past as I wrote that sentence. A herring gull, floating overhead on outstretched wings, completely unaware that what it was doing is as close to a miracle as makes no difference.

The air is big. Really big. Much bigger than the ground. To live with the ability to travel through it is to live a three-dimensional life. The rest of us, earthbound, flat-footed, leaden – we live in two dimensions. Flat lives, limited to north, south, east, west. From the human point of view, flight remains aspirational. It is a metaphor for freedom, of course, but it represents something else too. Because we can't do it. For all our ingenuity, our mammoth brains, our legendary problem-solving skills that are surely the envy of all other life forms on this planet, we can't do it. And while we have found ways round it, the true understanding of what it is, the sense of knowing that comes with innate ability, will always elude us. We work hard to unpick it, but even after all this time, and with all that human cleverness, we still barely understand how a fly flies. And when we try to emulate it, our efforts are lumbering when compared with the fruits of millions of years of evolution. As we remain earthbound, looking upwards in mute wonder, the dragonfly whizzes, the albatross soars, the bat jinks and darts in pursuit of the moth. And all we can do – must do – is look on and cheer.

ACKNOWLEDGEMENTS

Many people make a book, but only one gets their name on the front cover. This seems to me deeply unfair. This particular book would never have appeared without the help of many skilful and generous people.

Firstly, I owe a great deal to Simon Spanton, who has always been open to my wildest ideas, and without whom none of my last four books would have seen the light of day. Once the book was more or less written (or at least so I thought), Sarah Rigby and Pippa Crane were on hand to put me right. They guided me through the editing process with extraordinary skill and patience, understanding not just what was wrong with it, but – even more importantly – suggesting how I might mend it. Authorly instincts often resist such interference – it helps to remember that they are always right. Copyeditor Jill Burrows is by now used to my bad habits, not least the repeated misplacement of the word 'only'. I thank her for her continued patience. The beautiful and striking cover was designed by Jo Walker. To all at Elliott & Thompson who guide a book from conception to publication and beyond: thank you.

For the technical aspects of a subject that seemed, like the universe, to keep on expanding, I relied on the accumulated knowledge of experts in the various fields. Alex Evans read the finished manuscript and pointed out several instances of vagueness, ambiguity, waffle and plain factual inaccuracy. Jo Ferguson and Lisa Worledge gave their time and bat knowledge freely and generously. I still know little about bats, but thanks to them I now know more than I used to. And whenever I think I know a lot about birds, a chat with David Darrell-Lambert puts me right. The informative and entertaining *Terrible Lizards* podcast was my introduction to the world of pterosaurs, so my gratitude goes to Iszi Lawrence and Dave Hone for producing it, and for making long walks round south London seem much shorter. If there are mistakes (there are bound to be), they are mine alone.

Laura Pritchard reads early drafts and, often by asking questions I can't answer, steers me towards the righteous path. David Durose nudged me into action at a crucial point in the 'writing' process with the sage words 'perhaps it's time to stop circling the wagons and actually write'. And Isabel Rogers knows the right things to say when I send her scrappy bits of nonsense.

As always, my wife Tessa provided invaluable and bottomless support, sympathy and encouragement. And, although neither of us realised it at the time, my son Oliver's passion for origami a few years ago, during which time he made and threw multiple paper planes, planted a seed.

Finally, a book doesn't need readers to exist, but knowing it might have at least a few of them does help the process along. So thank you for making it this far. And if I have forgotten anyone, consider yourself thanked.

NOTES

Chapter 1: The Mayfly

1. M. Engel, D. Grimaldi, 'New light shed on the oldest insect', *Nature*, vol. 427 (2004), pp. 627–630; https://doi.org/10.1038/nature02291
2. C. Haug, J. T. Haug, 'The presumed oldest flying insect: more likely a myriapod?', *PeerJ*, 5:e3402 (2017); https://doi.org/10.7717/peerj.3402
3. C. Brauckmann, J. Schneider, 'Ein unter-karbonisches Insekt aus dem Raum Bitterfeld/Delitzsch (Pterygota, Arnsbergium, Deutschland) [A Lower Carboniferous insect from the Bitterfeld/Delitzsch area (Pterygota, Arnsbergian, Germany)]', *Neues Jahrbuch für Geologie und Paläontologie*, vol. 1 (1996), pp. 17–30; https://doi.org/10.1127/njgpm/1996/1996/17
4. R. Fortey, *Life: An Unauthorised Biography* (HarperCollins, 1998)
5. D. E. Alexander, *On the Wing* (Oxford University Press, 2015)
6. S. E. Reynolds, 'Mayfly metamorphosis: Adult winged insects that molt', *Proceedings of the National Academy of Sciences USA*, vol. 118, no. 38 (2021); https://doi.org/10.1073/pnas.2114128118
7. I. Dawson, 'Five things you didn't know about insect wings', YouTube (2019); https://www.youtube.com/watch?v=exdL2H9PzJA

Chapter 2: The Dragonfly

1. N. Lane, *Oxygen: The Molecule that Made the World* (Oxford University Press, 2002)
2. H. Jurikova, M. Gutjahr, K. Wallmann et al., 'Permian–Triassic mass extinction pulses driven by major marine carbon cycle perturbations', *Nature Geoscience*, vol. 13 (2020), pp. 745–750; https://doi.org/10.1038/s41561-020-00646-4

3. C. C. Labandeira, 'The fossil record of insect extinction: New approaches and future directions', *American Entomologist*, vol. 51, no. 1 (2005), pp. 14–29; https://doi.org/10.1093/ae/51.1.14
4. A. Sverdrup-Thygeson, *Extraordinary Insects* (Simon & Schuster, 2020)
5. Dawson, 'Five things you didn't know about insect wings'
6. G. Glaeser, H. F. Paulus, W. Nachtigall, *The Evolution of Flight* (Springer International, 2017)
7. Q. Li, M. Zheng, T. Pan, et al., 'Experimental and numerical investigation on dragonfly wing and body motion during voluntary take-off', *Scientific Reports*, vol. 8, 1011 (2018); https://doi.org/10.1038/s41598-018-19237-w
8. D. E. Alexander, *Why Don't Jumbo Jets Flap Their Wings?* (Rutgers University Press, 2009)
9. A. T. Bode-Oke, S. Zeyghami, H. Dong, 'Flying in reverse: kinematics and aerodynamics of a dragonfly in backward free flight', *Journal of the Royal Society Interface*, vol. 15, no. 143 (2018); http://doi.org/10.1098/rsif.2018.0102
10. R. Futahashi, R. Kawahara-Miki, M. Kinoshita, K. Yoshitake, S. Yajima, K. Arikawa and T. Fukatsu, 'Extraordinary diversity of visual opsin genes in dragonflies', *Proceedings of the National Academy of Sciences*, vol. 112, no. 11 (2015); https://www.pnas.org/doi/full/10.1073/pnas.1424670112
11. S. D. Wiederman, J. M. Fabian, J. R. Dunbier, D. C. O'Carroll, 'A predictive focus of gain modulation encodes target trajectories in insect vision', *eLife*, 6:e26478 (2017); https://doi.org/10.7554/eLife.26478

Chapter 3: The Beetle

1. S. Q. Zhang, L. H. Che, Y. Li et al., 'Evolutionary history of Coleoptera revealed by extensive sampling of genes and species', *Nature Communications*, vol. 9, 205 (2018); https://doi.org/10.1038/s41467-017-02644-4
2. R. Jones, *Beetles* (Collins New Naturalist Library, 2018)
3. D. Linz, A. Hu, M. Sitvarin, Y. Tomoyasu, 'Functional value of elytra under various stresses in the red flour beetle, *Tribolium castaneum*', *Scientific Reports*, vol. 6, 34813 (2016); https://doi.org/10.1038/srep34813
4. K. Saito, S. Nomura, S. Yamamoto, R. Niiyama, and Y. Okabe, 'Investigation of hindwing folding in ladybird beetles by artificial elytron transplantation and microcomputed tomography', *Proceedings of the National Academy of Sciences*, vol. 114, no. 22 (2017); https://doi.org/10.1073/pnas.1620612114

5. H. V. Phan, H. C. Park, 'Mechanisms of collision recovery in flying beetles and flapping-wing robots', *Science*, vol. 370, no. 652 (2020); https://doi.org/10.1126/science.abd3285

6. T. L. Erwin, 'Tropical forests: Their richness in Coleoptera and other arthropod species', *The Coleopterists Bulletin*, vol. 36, no. 1 (1982); https://www.jstor.org/stable/4007977

Chapter 4: The Fly

1. R. Rader, S. A. Cunningham, B. G. Howlett, D. W. Inouye, 'Non-bee insects as visitors and pollinators of crops: Biology, ecology, and management', *Annual Review of Entymology*, vol. 65 (2020), pp. 391–407; https://doi.org/10.1146/annurev-ento-011019-025055

2. E. McAlister, *The Secret Life of Flies* (The Natural History Museum, 2017)

3. M. Dickinson, 'How flies fly', YouTube (2020); https://www.youtube.com/watch?v=lv5vDW59hbY

4. S. E. Farisenkov, D. Kolomenskiy, P. N. Petrov et al., 'Novel flight style and light wings boost flight performance of tiny beetles', *Nature*, vol. 602 (2022), pp. 96–100; https://doi.org/10.1038/s41586-021-04303-7

5. T. Weis-Fogh, 'Quick estimates of flight fitness in hovering animals, including novel mechanisms for lift production', *Journal of Experimental Biology*, vol. 59, no. 1 (1973), pp. 169–230; https://doi.org/10.1242/jeb.59.1.169

6. F. van Breugel, M. Dickinson, 'Plume-tracking behavior of flying *Drosophila* emerges from a set of distinct sensory-motor reflexes', *Current Biology*, vol. 24, no. 3 (2014), pp. 274–286; https://doi.org/10.1016/j.cub.2013.12.023

Chapter 5: The Bee

1. D. L. Altshuler, W. B. Dickson, J. T. Vance, S. P. Roberts, M. H. Dickinson, 'Short-amplitude high-frequency wing strokes determine the aerodynamics of honeybee flight', *Proceedings of the National Academy of Sciences*, vol. 102, no. 50 (2005), pp. 18213–18218; https://doi.org/10.1073/pnas.0506590102

2. J. M. Peters, N. Gravish, S. A. Combes, 'Wings as impellers: Honey bees co-opt flight system to induce nest ventilation and disperse pheromones', *Journal of Experimental Biology*, vol. 220, no. 12 (2017), pp. 2203–2209; https://doi.org/10.1242/jeb.149476

Chapter 6: The Butterfly

1. C. Stefanescu, F. Páramo, S. Åkesson et al., 'Multi-generational long-distance migration of insects: Studying the painted lady butterfly in the Western Palaearctic', *Ecography*, vol. 36 (2013), pp. 474–486; https://doi.org/10.1111/j.1600-0587.2012.07738.x
2. University of Pittsburgh, 'Tiny sensor used to track the migratory patterns of monarch butterflies', *ScienceDaily* (2022); www.sciencedaily.com/releases/2022/05/220502120508.htm
3. H. Ghiradella, 'Structure of butterfly scales: Patterning in an insect cuticle', *Microscopy Research & Technique*, vol. 2, no. 5 (1994), pp. 429–438; https://doi.org/10.1002/jemt.1070270509
4. E. Osterloff, 'Why do some butterflies and moths have eyespots?'; https://www.nhm.ac.uk/discover/why-do-butterflies-have-eyespots.html
5. T. J. B. van Eldijk, T. Wappler, P. K. Strother et al., 'A Triassic-Jurassic window into the evolution of Lepidoptera', *Science Advances*, vol. 4, no. 1 (2018); https://doi.org/10.1126/sciadv.1701568
6. L. C. Johansson, P. Henningsson, 'Butterflies fly using efficient propulsive clap mechanism owing to flexible wings', *Journal of the Royal Society Interface*, vol. 18, no. 174 (2021); http://doi.org/10.1098/rsif.2020.0854
7. *David Attenborough's Conquest of the Skies*, Colossus Productions, 2015

Chapter 7: The Pterosaur

1. M. T. Wilkinson, D. M. Unwin, C. P. Ellington, 'High lift function of the pteroid bone and forewing of pterosaurs', *Proceedings of the Royal Society Biological Sciences*, vol. 273, no. 1582 (2005); https://doi.org/10.1098/rspb.2005.3278
2. I. Lawrence, Dr. D. Hone, 'Pterodactylus', *Terrible Lizards* [podcast], season 7, episode 1 (1 June 2022); https://terriblelizards.libsyn.com
3. M. Habib, 'Comparative evidence for quadrupedal launch in pterosaurs', *Zitteliana Reihe B*, vol. 28 (2008); https://www.researchgate.net/publication/49185735
4. M. Witton, 'Raising the curtain on a new pterosaur' (2009); https://www.fleshandstone.net/healthandsciencenews/1652.html
5. I. Lawrence, Dr. D. Hone, 'Anurognathids', *Terrible Lizards* [podcast], season 7, episode 4 (22 June 2022); https://terriblelizards.libsyn.com

6. M. P. Witton, *Pterosaurs: Natural History, Evolution, Anatomy* (Princeton University Press, 2013)
7. Lawrence, Hone, 'Anurognathids', *Terrible Lizards* [podcast]
8. M. A. Brown, K. Padian, Preface, *Journal of Vertebrate Paleontology*, 41: sup1, 1 (2021); http://doi.org/10.1080/02724634.2020.1853560
9. Witton, *Pterosaurs*

Chapter 8: The Archaeopteryx

1. R. O. Prum, 'Development and evolutionary origin of feathers,' *Journal of Experimental Biology*, vol. 285, no. 4 (2002); https://doi.org/10.1002/(SICI)1097-010X(19991215)285:4<291::AID-JEZ1>3.0.CO;2-9
2. K. P. Dial, 'Wing-assisted incline running and the evolution of flight', *Science*, vol. 299, no. 5605 (2003); http://doi.org/10.1126/science.1078237

Chapter 9: The Penguin

1. C. L. Williams, J. C. Hagelin, G. L. Kooyman, 'Hidden keys to survival: the type, density, pattern and functional role of emperor penguin body feathers', *Proceedings of the Royal Society Biological Sciences*, vol. 282, no. 1817 (2015); https://doi.org/10.1098/rspb.2015.2033
2. T. Hanson, *Feathers: The Evolution of a Natural Miracle* (Basic Books, 2011)

Chapter 10: The Goose

1. G. R. Scott, L. A. Hawkes, P. B. Frappell et al., 'How bar-headed geese fly over the Himalayas', *Physiology*, vol. 30, no. 2 (2015), pp. 107–15; http://doi.org/10.1152/physiol.00050.2014
2. R. Wiltschko, I. Schiffner, P. Fuhrmann, W. Wiltschko, 'The role of the magnetite-based receptors in the beak in pigeon homing', *Current Biology*, vol. 20, no. 17 (2010); https://doi.org/10.1016/j.cub.2010.06.073
3. J. Xu, L. E. Jarocha, T. Zollitsch et al., 'Magnetic sensitivity of cryptochrome 4 from a migratory songbird', *Nature*, vol. 594 (2021), pp. 535–540; https://doi.org/10.1038/s41586-021-03618-9
4. D. Naish, M. P. Witton, E. Martin-Silverstone, 'Powered flight in hatchling pterosaurs: evidence from wing form and bone strength', *Scientific Reports*, 11, 13130 (2021); https://doi.org/10.1038/s41598-021-92499-z

Chapter 13: The Pigeon

1. S. Potier, M. Lieuvin, M. Pfaff, A. Kelber, 'How fast can raptors see?', *Journal of Experimental Biology*, vol. 223, no. 1 (2020); https://doi.org/10.1242/jeb.209031
2. T. H. Worthy, S. J. Hand, J. P. Worthy, A. J. D. Tennyson, R. P. Scofield, 'A large fruit pigeon (Columbidae) from the Early Miocene of New Zealand', *The Auk*, vol. 126, no. 3 (2009), pp. 649–656; https://doi.org/10.1525/auk.2009.08244

Chapter 14: The Bat

1. G. Jones, M. W. Holderied, 'Bat echolocation calls: adaptation and convergent evolution', *Proceedings of the Royal Society Biological Sciences*, vol. 274, no. 1612 (2007); http://doi.org/10.1098/rspb.2006.0200
2. L. I. Amador, N. B. Simmons, N. P. Giannini, 'Aerodynamic reconstruction of the primitive fossil bat *Onychonycteris finneyi*', *Biology Letters*, vol. 15, no. 3 (2019); http://doi.org/10.1098/rsbl.2018.0857
3. G. F. Gunnell, N. B Simmons, 'Fossil evidence and the origin of bats', *Journal of Mammalian Evolution*, vol. 12 (2005), pp. 209–246; http://doi.org/10.1007/s10914-005-6945-2
4. D. Jebb, Z. Huang, M. Pippel et al., 'Six reference-quality genomes reveal evolution of bat adaptations', *Nature*, vol. 583 (2020), pp. 578–584; https://doi.org/10.1038/s41586-020-2486-3
5. J. Fullard, J. Dawson, 'The echolocation calls of the spotted bat Euderma maculatum are relatively inaudible to moths', *Journal of Experimental Biology*, vol. 200, no. 1 (1997), pp. 129–137; https://doi.org/10.1242/jeb.200.1.129
6. M. B. Fenton, G. P. Bell, 'Recognition of species of insectivorous bats by their echolocation calls, *Journal of Mammalogy*, vol. 62, no. 2 (1981), pp. 233–243; https://doi.org/10.2307/1380701
7. S. M. M. Brinkløv, L. Jakobsen, L. A. Miller, 'Echolocation in bats, odontocetes, birds, and insectivores', *Exploring Animal Behavior Through Sound: Volume 1*, edited by C. Erbe, J. A. Thomas (Springer, 2022); https://doi.org/10.1007/978-3-030-97540-1_12
8. R. Simon, M. W. Holderied, C. U. Koch, O. von Helversen, 'Floral acoustics: Conspicuous echoes of a dish-shaped leaf attract bat pollinators', *Science*, 333 (6042) (2011), pp. 631–633; https://doi.org/10.1126/science.1204210

9. J. R. Barber, D. Plotkin, J. J. Rubin et al., 'Anti-bat ultrasound production in moths is globally and phylogenetically widespread', *Proceedings of the National Academy of Sciences USA*, vol. 119, no. 25 (2022); https://doi.org/10.1073/pnas.2117485119
10. C. Tianxin, J-P Jin, 'Evolution of flight muscle contractility and energetic efficiency', *Frontiers in Physiology*, vol. 11 (2020); https://doi.org/10.3389/fphys.2020.01038
11. Y. Yongliang, G. Ziwu, 'Learning from bat: Aerodynamics of actively morphing wing', *Theoretical and Applied Mechanics Letters*, vol. 5, no. 1 (2015); https://doi.org/10.1016/j.taml.2015.01.009
12. S. Sterbing-D'Angelo et al., 'Bat wing sensors support flight control', *Proceedings of the National Academy of Sciences USA*, vol. 108, no. 27 (2011); https://doi.org/10.1073/pnas.1018740108
13. S. Sterbing-D'Angelo, M. Chadha, K. L. Marshall, C. F. Moss, 'Functional role of airflow-sensing hairs on the bat wing', *Journal of Neurophysiology*, vol. 117, no. 2 (2017), pp. 705–712; https://doi.org/10.1152/jn.00261.2016
14. M. Brock Fenton, N. B. Simmons, *Bats: A World of Science and Mystery* (University of Chicago Press, 2015)
15. U. Norberg, J. Rayner, 'Ecological morphology and flight in bats (Mammalia; Chiroptera): Wing adaptations, flight performance, foraging strategy and echolocation, *Royal Society of London Philosophical Transactions Series B*, vol. 316. (1987); https://doi.org/10.1098/rstb.1987.0030
16. Fenton & Simmons, *Bats*
17. Ibid.

SELECTED BIBLIOGRAPHY

Alexander, D. E., *On the Wing* (Oxford University Press, 2015)

Alexander, D. E., *Why Don't Jumbo Jets Flap Their Wings?* (Rutgers University Press, 2009)

Benito, J., Olivé, R., *Birds of the Mesozoic* (Lynx, 2022)

Brusatte, S., *The Rise and Fall of the Dinosaurs* (Macmillan, 2018)

Chandler, D., Cham, S., *Dragonfly* (New Holland Publishers, 2013)

Dalton, S., *Borne on the Wind* (Chatto & Windus, 1975)

Dalton, S., *The Miracle of Flight* (Merrell, 2001)

Dawkins, R., *Flights of Fancy* (Head of Zeus, 2021)

Dunn, J., *The Glitter in the Green* (Bloomsbury, 2021)

Fenton, M. B., Simmons, N. B., *Bats: A World of Science and Mystery* (University of Chicago Press, 2014)

Fortey, R., *Life: An Unauthorised Biography* (Flamingo, 1998)

Glaeser, G., Paulus, H., Nachtigall, W., *The Evolution of Flight* (Springer International, 2017)

Hanson, T., *Feathers: The Evolution of a Natural Miracle* (Basic Books, 2011)

Jones, R., *Beetles* (William Collins, 2018)

Martyniuk, M. P., *A Field Guide to Mesozoic Birds and Other Winged Dinosaurs* (Pan Aves, 2012)

McAlister, E., *The Secret Life of Flies* (Natural History Museum, 2017)

Newton, I., *Bird Migration* (William Collins, 2010)

Nicolson, A., *The Seabird's Cry* (William Collins, 2017)

Paul, G. S., *The Princeton Field Guide to Pterosaurs* (Princeton University Press, 2022)

Pickrell, J., *Flying Dinosaurs: How Fearsome Reptiles Became Birds* (New South Publishing, 2014)

Reilly, J., *The Ascent of Birds* (Pelagic Publishing, 2018)

Shaw, S. R., *Planet of the Bugs* (The University of Chicago, 2014)

Sverdrup-Thygeson, A., *Extraordinary Insects* (Mudlark, 2019)

Tennekes, H., *The Simple Science of Flight* (Massachusetts Institute of Technology, 2009)

Videler, J., *Avian Flight* (Oxford University Press, 2005)

Weidensaul, S., *A World on the Wing* (W. W. Norton, 2021)

Wellnhofer, P., *The Illustrated Encyclopedia of Pterosaurs* (Salamander, 1991)

Witton, M. P., *Pterosaurs* (Princeton University Press, 2013)

INDEX